HUMAN GENETICS NOTES

M. M. GREEN

Department of Genetics
University of California, Davis

HUMAN GENETICS NOTES

ADDISON-WESLEY PUBLISHING COMPANY
Reading, Massachusetts
Menlo Park, California · London · Amsterdam · Don Mills, Ontario · Sydney

Reproduced by Addison-Wesley from camera-ready copy prepared by the author.

ISBN 0-201-02599-X
ABCDEFGHIJ-AL-798765

Preface

This syllabus is not a textbook of Human Genetics nor is it meant to substitute for one. Rather it is intended to supplement a quarter or semester course of approximately 40 lectures in Human Genetics for students of Human Biology whose preparation in biology, statistics and biochemistry is minimal. In scope, this course is designed to provide students with an understanding of basic genetic principles with emphasis on man. Genetic entities of medical importance are considered within the context of Human Genetics but there is no manifest stress on Medical Genetics. Where possible, emphasis is placed on the ideas of genetics rather than the facts of genetics.

Why a syllabus? Experience teaches that if the lecture method of university instruction is to be effective, some means must be found to minimize the amount of note-taking and to maximize the amount of listening done by students. This is the primary purpose of the syllabus. The subject matter covered is sufficiently detailed so that a student need add only occasional supplementary notes to the lecture presentation. This, of course, presumes that the "meat" of the lectures is built around the subject matter presented in the syllabus. Detailed lectures plus the syllabus have successfully obviated the need for a textbook.

In using the syllabus, it is assumed that having completed the prerequisite courses (a first course in Biology and an elementary course in Statistics), students are familiar with the fundamental concepts of cell division, mitosis and meiosis, the basic ideas of Mendelian genetics and the elementary notions of probability. In this connection perusal of section 2, "A basic vocabulary" may be instructive. For students in need of review sections 3 and 4 have been included. Problems covering each section are also included to allow students to check on their mastery of the subject.

For more detailed discussions of topics in Human and Medical Genetics the following books should be consulted.

Stern, Curt. 1973. Principles of Human Genetics. 3rd ed. 891 pp. W. H. Freeman and Co., San Francisco and London.

Whittinghill, M. 1965. Human Genetics and its Foundations. 431 pp. Reinhold Publishing Corp., New York.

Levitan, Max and Ashley Montagu. 1971. Text book of Human Genetics. 922 pp. Oxford University Press, New York.

McKusick, V. A. 1969. Human Genetics. 2nd ed. 221 pp. Prentice-Hall, Inc., Englewood Cliffs, NJ.

McKusick, V. A. and R. Claiborne, editors. Medical Genetics. 300 pp. H. P. Publishing Co., New York.

Stevenson, A. C. and B. C. C. Davison with M. W. Oakes. 1970. Genetic Counselling. 355 pp. J. B. Lippincott Co., Philadelphia.

Roberts, J. A. F. 1970. An Introduction to Medical Genetics. 5th ed. 296 pp. Oxford University Press, London.

Harris, H. 1970. The Principles of Human Biochemical Genetics. 328 pp. American Elsevier Publishing Co., Inc., New York.

Table of Contents

Section

"If we enable the weak and the deformed to live and to propogate their kind, we face the prospect of a genetic twilight. But if we let them die or suffer when we can save or help them, we face the certainty of a moral twilight."

Th. Dobzhansky

1. Introduction

Of all the organisms inhabiting the earth, the one which intrigues, perplexes and exasperates man more than any other is man himself. It is, therefore, not altogether unexpected that for the student of biology the study of all aspects of the biology of man has been especially attractive and challenging from the very beginning of the science. Nonetheless, it is somewhat surprising that until comparatively recently man as the object of genetic investigation has lagged behind genetic investigations of many other organisms. Generation time, small family size, uncontrolled matings - all have contributed to discouraging genetic studies of man. With the development of several genetic, cytological and biochemical techniques, the aforementioned difficulties have been in a large measure overcome. Now the formal genetics of man can be studied with almost the same facility, precision and rigor as is the genetics of such laboratory organisms as bacteriophage T4, *Escherichia coli*, *Drosophila melanogaster* and the mouse. A further impetus to the study of genetics of man stems from the recognition of a large number of human diseases whose etiology has a genetic or chromosomal basis. The advent of the atomic age with the detonation of the Hiroshima bomb has also had its impact by emphasizing the probable genetic damage to man which can come from exposure to high energy radiation. More recently, the provocative test-tube manipulations of DNA have posed the issue of the social implications of the potential rearrangement of the genetic material of a variety of organisms including man. Thus, it has become increasingly clear that both a comprehension and an appreciation of the principles of human genetics are necessary components in the decision making processes upon which the future social welfare, perhaps survival, of mankind will depend.

In the presentation which follows, where possible, a simple rationale has been adopted. The optimal circumstance in the development of a basic genetic principle or concept is considered first. Often this involves collection of cogent facts or critical experiments with specific experimental animals, plants or microbes. This is followed by the parallel genetic event in man, not infrequently a suboptimal situation.

NOTES

2. A Basic Vocabulary

In preparation for the study of Mendelian genetics of man, a review of the basic facts of mitosis and meiosis and of Mendel's first law, the law of segregation, is in order. Especially important is a review of the various terms commonly employed for without a proper vocabulary, comprehension is made most difficult. The following list is a synopsis of the commonly used terms.

Mitosis and meiosis:

metabolic cell	centromere	daughter cells	polar body
nucleus	chromatid	disjunction	secondary spermatocyte
nucleolus	prophase	oogonium	spermatid
cytoplasm	metaphase	spermatogonium	spermiogenesis
chromosome	anaphase	primary oocyte	ovum
autosome	telophase	primary spermatocyte	spermatozoon
sex chromosome	spindle fiber	synapsis	fertilization
X chromosome	equatorial plate	tetrad	zygote
Y chromosome	cytokinesis	secondary oocyte	karyotype

Mendelian genetics:

phenotype	recessive	P generation	heterozygous
genotype	codominant	F_1 generation	test cross
contrasting traits	gene	F_2 generation	monohybrid
dominant	allele	homozygous	dihybrid

3. Basic Cytology--A Synopsis

Modern genetics and cytology have their origins in the mid 19th and early 20th centuries. During this period the basic precepts of cytology and genetics gradually unfolded, more or less independently, culminating in the chromosomal principle of hereditary transmission as we recognize it today. This principle, simply stated, holds that the genes which endow individuals with their unique hereditary features are an integral part of the chromosomes of each cell. In multicellular organisms sexual reproduction mediates the transmission of chromosomes (and therefore genes) from one generation to the next.

NOTES

The chronology of events climaxing in the chromosomal principle begins in 1838-9 with the announcement of the cell theory. According to this theory all living higher organisms are aggregates of fundamental units called cells. Cells do not arise de novo but derive only from pre-existing cells. Subsequently the microscopic study of cells revealed the relative structural complexity of cells. Such study also demonstrated that per individual plant or animal species there occur specific cellular components which distinguish the cells of one species from those of other species. Those microscopically discernable cellular components which invariably distinguish the cells of one species from the cells of other species are the chromosomes which occur in the cellular nuclei of living organisms.

The chromosomes of one species can be distinguished from those of another species according to a number of criteria including (1) their total number or chromosome complement and (2) the characteristic morphology of the individual chromosomes. Chromosomes are often best studied during mitosis, the process whereby somatic cells divide. Such studies reveal that the chromosomes of each species occur as a fixed number of homologous pairs, the individual members of a pair being morphologically identical but often distinguishable from other (nonhomologous) chromosome pairs. Thus the total number of chromosomes or chromosome complement is said to be diploid (or 2n). Chromosomes of one pair often can be distinguished from any other pair by their overall lengths or by the location of the centromere. Where the centromere is median in location, the chromosome consists of two arms of equal length. Such chromosomes are called metacentric. Where the centromere is submedian, the arm lengths are unequal and such chromosomes are said to be submetacentric. Where the centromere is located very close to one end, the chromosome is said to be acrocentric. By including all criteria--total number, overall length, centromere location and some special morphological features e.g. satellites, constrictions, nucleolus organizer--the chromosome complement or karyotype of one species is objectively distinguishable from that of other species.

With many sexually reproducing organisms, particularly animals, it is often possible to distinguish between males and females on the basis of karyotype. In males of many animal species all but one pair of homologous chromosomes occur as identical (homomorphic) pairs. The chromosomes of one pair are morphologically distinctive or heteromorphic. They are designated X and Y and together are called the sex chromosomes. In females of the same

NOTES

species the chromosomes consist only of homologous pairs including a sex chromosome set of two X chromosomes. Chromosomes not distinguishable on the basis of sex are collectively called the autosomes.

Although in many animal species, e.g. man, the male is XY and the female XX, in some animal species, especially birds and butterflies, the female is XY and the male XX. There are some animal species where the sexes cannot be separated on the basis of karyotype.

Since the somatic or body cells of an animal have identical chromosome complements irrespective of function, it is reasonable to inquire how this constancy is brought about. The process of somatic cell division or mitosis guarantees the regular distribution of chromosomes from parental cell to daughter cells in the following way.

The transition from a metabolic (interphase) cell to a dividing cell is accompanied by several characteristic changes in the cell's nucleus. The metabolic cell typically includes a nucleus bounded by a nuclear membrane and a distinct nucleolus. Chromosomes as such are not visible. The onset of mitosis is presaged by the appearance of the chromosomes which have already reduplicated as long, attenuated threads, by the gradual dissolution of the nucleolus and by the progressive disappearance of the nuclear membrane. This is the prophase stage of mitosis (diagram 1a and 1b). By the end of prophase the chromosomes have condensed maximally and the spindle is formed to which the centromeres of each chromosome are attached by the spindle fibers (diagram 1c). The metaphase stage is characterized by the alignment of the centromeres at the spindle equator forming the equatorial plate (diagram 1d). At anaphase the reduplicated or sister chromosomes separate and move to opposite poles of the cell. Thereby each daughter cell is assured of quantitatively and qualitatively the same chromosomes as the parental cell (diagram 1e). The final or telophase stage of mitosis is, in effect, the reverse of prophase. The chromosomes decondense and become less and less distinct, the nuclear membrane reappears as does the nucleolus (diagram 1f). With the division of the cytoplasm (cytokinesis) two daughter cells are formed whose chromosome complements are identical in all respects to one another and to the parental cell from which they arose.

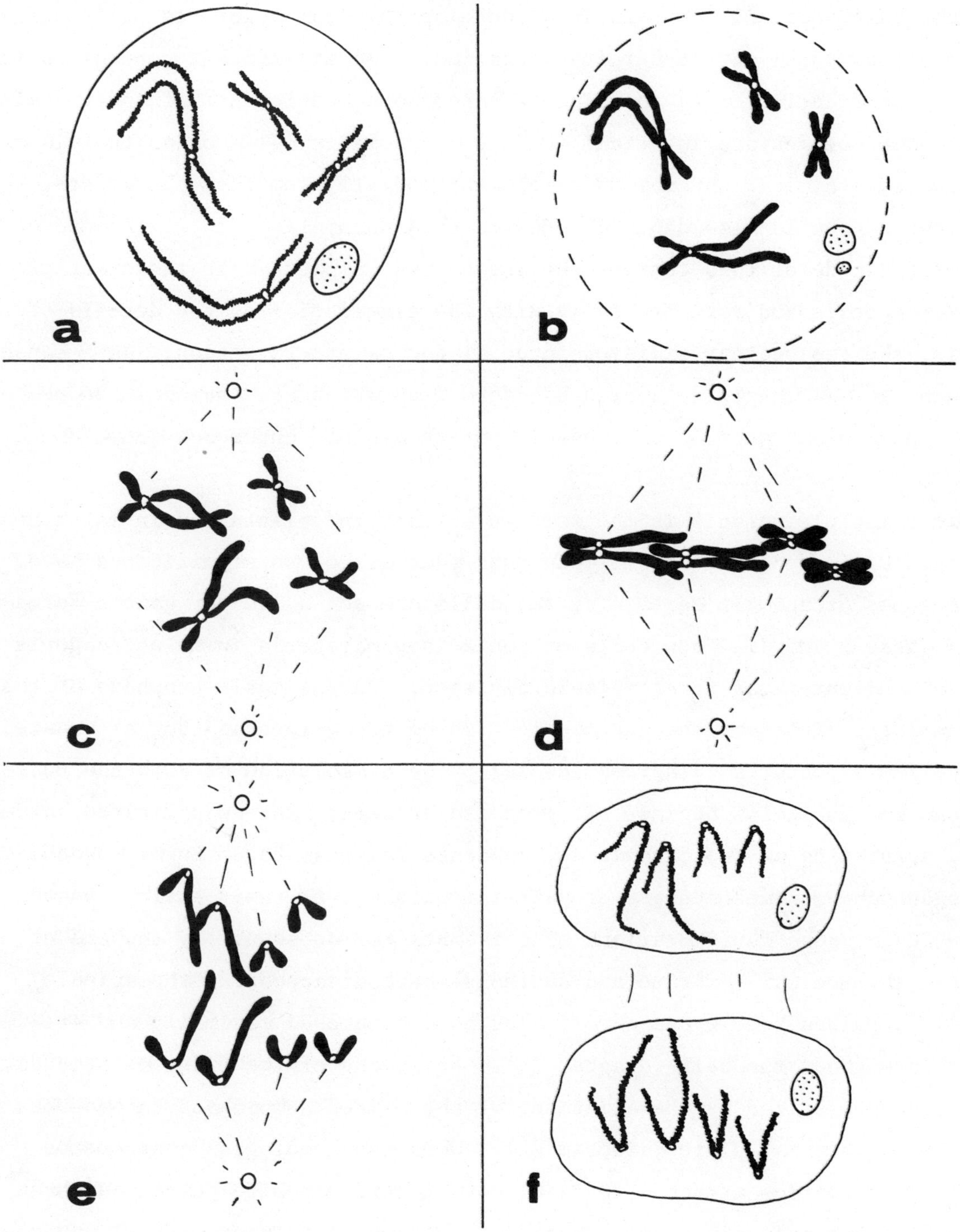

Plate 1. Diagrammatic representation of mitosis in an animal with two pairs of chromosomes.

By the end of the 19th century the general principles of animal reproduction had been unraveled. Fertilization of ovum by sperm to produce the zygote had been described and the details of development from zygote to fully differentiated individual were under investigation. Two significant generalizations needed clarification: (1) the diploid chromosome complements of individuals of the same species are identical and (2) all somatic cells of an individual have the same diploid chromosome complement and all stem from the zygote. Ergo, the zygote is also diploid. How is this accomplished? By the end of the first decade of this century the answer was complete with the description of gametogenesis and more precisely with the elucidation of the details of meiosis, the two nuclear divisions crucial for gamete formation. As a consequence of meiosis, haploid gametes are formed in which are included one member of each homologous pair of chromosomes in the diploid chromosome complement.

In principle meiosis is the same in animals and plants and in males and females. Differences in detail occur but they are of no significance here. In the gonad of animals certain of the cells are set aside for gamete formation. These primordial sex cells or gonia (spermatogenia in males, oogonia in females) and enter the first meiotic division. In the early prophase of the first meiotic division, the chromosomes can be recognized as long attenuated thread-like structures (diagrams 2a, 2b). The dissolution of both the nuclear membrane and nucleolus begins. As prophase proceeds, the reduplicated chromosomes, consisting of two chromatids, condense (diagram 2c). By mid-prophase homologous chromosomes synapse or pair intimately over their entire lengths figure 2d). Cross configurations or chiasmata can be seen. By the end of prophase the nuclear membrane and nucleolus have disappeared, the spindle forms, the paired homologous chromosomes have condensed maximally and are well spread throughout the cell (diagram 2e). Metaphase of the first meiotic division is characterized by the alignment of the paired homologous chromosomes along the equatorial plate (diagram 2f). When the homologous centromeres (with two chromatids attached to each) move toward opposite poles, anaphase of the first meiotic division has begun (diagram 2g). After maximal movement, the primary gametocyte contains two sets of chromosomes at each pole, each with half the number of centromeres of the parental cell (diagram 2h). Telophase of the first meiotic division follows with the formation of two daughter secondary gametocytes (secondary spermatocytes in males and a secondary oocyte and a polar body in females). In many animals the second meiotic division proceeds almost immediately without an interphase (metabolic) nucleus

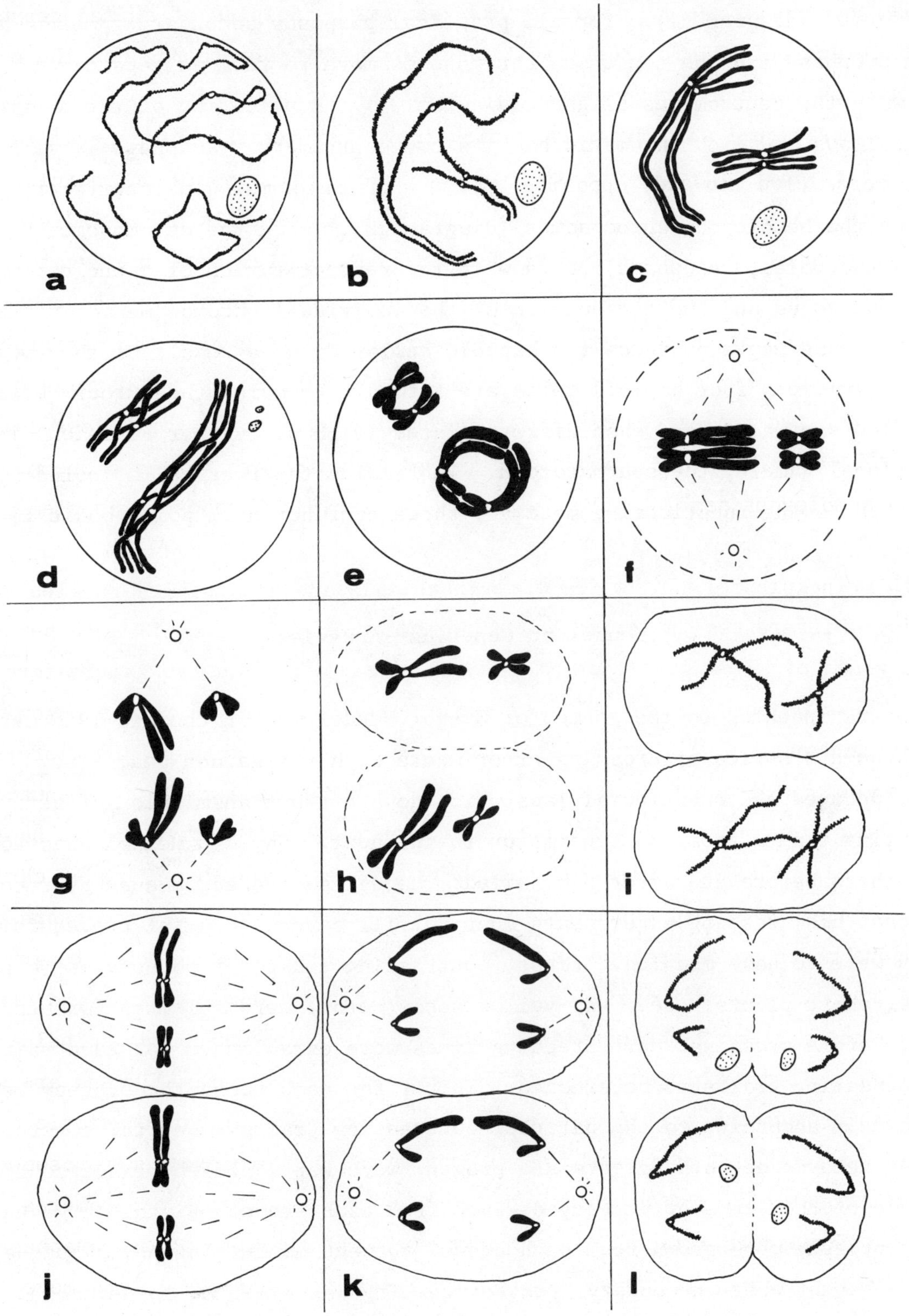

Plate 2. Diagrammatic representation of meiosis in an animal with two pairs of chromosomes.

being formed (diagram 2i). For all practical purposes secondary gametocytes are in prophase and the condensed chromosomes move to form an equatorial plate with the centromeres aligned (diagram 2j). Replication of the centromeres follows and each chromatid has its own centromere. Anaphase is marked by the separation towards opposite poles of the homologous centromeres and thus the homologous chromosomes (diagram 2k). When the chromosomes have reached the poles, telophase occurs with the reconstitution of a nuclear membrane, nucleolus and the attenuation of the individual chromosomes. Thus each secondary gametocyte produces two haploid cells, or as a result of the two meiotic divisions four haploid cells are formed. In males these four cells are called spermatids and each differentiates (without further division) into a functional gamete, the spermatozoan. In females the four cells include one ovum which is the functional gamete and three nonfunctional polar bodies.

4. Mendelian Genetics--A Review

We are indebted to the genius of Gregor Mendel who in the mid 19th century through a series of crossing experiments with the garden pea, *Pisum sativa*, deduced the fundamental laws which govern the transmission of heritable contrasting traits from one generation to the next. Why did Mendel succeed where others before and after him failed? Mendel succeeded because he recognized that his crossing experiments must fulfill several crucial prerequisites if they were to have meaning. First, contrasting traits or phenotypes (e.g. tall vs. short plants; green vs. yellow seeds, etc.) were selected for study. Second, before crossing, pure breeding lines were established for each pair of the contrasting traits to be studied. Third, for each cross the progeny were tallied only according to the parental contrasting traits. Fourth, careful pedigree records of both parents and progeny were kept. Fifth, exact counts were made of the kinds of progeny derived from each cross. Sixth, large numbers of progeny were grown from each cross, thereby minimizing any sampling bias. Seventh, where necessary, progeny testing was employed to determine the genetic constitution or genotype of individuals.

Mendel's first law or the law of segregation is here illustrated by crosses between mice with the contrasting coat phenotypes self-colored (or wild-type) vs. piebald spotting. Piebald spotted mice were isolated from a wild population and a pure line established. Thus from the cross spotted X

NOTES

spotted the progeny are invariably spotted. A pure self-colored line is also available in which the cross self X self invariably results in self-colored progeny. Crosses between the two lines and their progeny gave the following results:

P ♀ self X ♂ spotted or ♂ self X ♀ spotted

F_1 all self

F_1 X F_1 → F_2 803 self and 269 spotted (total 1072).

Note: the frequencies of the F_2 phenotypes approximate 3 self: 1 spotted. Crosses of F_1 self-colored individuals X spotted individuals resulted in 269 self and 263 spotted progeny or 1:1. Crosses of F_2 self-colored progeny to spotted individuals (a test cross!) distinguished two kinds of individuals. Some crosses (ca. 1/3) with F_2 self-colored individuals were indistinguishable from the P cross: thus all the progeny were self-colored. Other crosses (ca. 2/3) with F_2 self-colored individuals were indistinguishable from the cross of F_1 X spotted: thus the progeny fell into two classes: ca. 1/2 self-colored and ca. 1/2 spotted. Crosses of F_2 spotted X F_2 spotted and F_2 spotted X pure line spotted produced only spotted progeny.

In Mendelian terms the results of the foregoing crosses are interpreted as follows. The phenotypes self and spotted are conditioned by the hereditary determiners or genes denoted _S_ and _s_, respectively. These genes are present in duplicate (_SS_ vs. _ss_) or are homozygous in the somatic cells of the pure line individuals. During gametogenesis each member of a gene pair has an equal probability of being included in a gamete. An _SS_ individual produces only _S_ gametes and an _ss_ individual only _s_ gametes. In the cross _SS_ X _ss_, a gamete carrying _S_ unites with a gamete carrying _s_ to produce a zygote, _Ss_, which differentiates into a self-colored individual. Each F_1 individual is _Ss_ or heterozygous in genotype because following gametogenesis each produces gametes of two types in equal frequency, 1/2 _S_ and 1/2 _s_. In an F_1 X F_1 cross the random union of the gametes produced by each F_1 individual leads to the expected zygotic frequencies of 1/4 _SS_, 2/4 _Ss_, 1/4 _ss_. Since the F_1 self-colored individuals (_Ss_) cannot be phenotypically discriminated from the P self-colored individuals (_SS_), the phenotypes expected among the F_2 individuals are 3/4 self, 1/4 spotted. (Therefore _S_ is said to be dominant to _s_, or _s_ is recessive to _S_). It will be noted further that among the F_2 self-colored individuals, 1/3 are expected to be homozygous or _SS_ and 2/3 are

NOTES

expected to be heterozygous or Ss. Precisely these results were obtained when the F_2 self-colored individuals were test-crossed.

Because in heterozygous individuals the genes S and s invariably segregate to separate gametes, they are called allelic genes. Accordingly Mendel's first law or the law of segregation may be defined as follows: the genes of each allelic pair are segregated to separate gametes, each gene being unaffected by the other.

Note that the segregation of allelic genes conforms precisely to the disjunction of homologous autosomes during meiosis. Thus Mendel's first law applies specifically to genes localized (linked!) to the autosomal chromosomes. This association will be proved subsequently.

Mendel's second law or the law of independent assortment governs the concurrent inheritance of two or more distinctive pairs of contrasting traits. Each pair of allelic genes is transmitted in association with a specific pair of autosomes. This is illustrated by studying the joint inheritance of self vs. spotted and agouti vs. black coat color. (Agouti and black, conditioned by the alleles A vs. a, exhibit conventional Mendelian segregation.) A cross is made between homozygous spotted, agouti and homozygous self, black mice. The F_1 individuals are all self, agouti in phenotype. A series of F_1 X F_1 crosses produced the following F_2 progeny: 301 self, agouti; 102 self, black; 95 spotted agouti and 30 spotted, black (total = 528). Are these results compatible with the independent assortment of the two sets of contrasting traits? The F_2 phenotypes expected in the case of independent assortment can be computed as follows. Symbolically the P cross of spotted, agouti X self, black can be written ss;AA X SS;aa with the F_1 all Ss;Aa or self, agouti in phenotype. An F_1 X F_1 cross or Ss;Aa X Ss;Aa represents two segregation crosses occurring simultaneously but independently. One cross is Ss X Ss from which the expected phenotypes of the progeny are 3/4 self and 1/4 spotted. The second cross is Aa X Aa from which the expected phenotypes are 3/4 agouti and 1/4 black. What is the combined phenotypic expectation in the F_2? Since the two segregations are independent, their combined or joint phenotypic expectations are the product of their individual frequencies or

3/4 self + 1/4 spotted
x 3/4 agouti + 1/4 black

9/16 self, agouti + 3/16 self, black + 3/16 spotted, agouti + 1/16 spotted black.

NOTES

This is the classical 9:3:3:1 Mendelian phenotypic ratio for independently assorting traits. Among 528 F_2 progeny the expected phenotypes are computed as: self, agouti = 9/16 X 528 = 297; self, black = 3/16 X 528 = 99; spotted, agouti = 3/16 X 528 = 99; and spotted, black = 1/16 X 528 = 33. Clearly the agreement between observation and expectation is excellent and the two results do not differ statistically.

Note that the independent assortment of two pairs of allelic genes parallels precisely the independent disjunction of two pairs of autosomes during meiosis.

The genotypes expected among the F_2 progeny in the aforementioned cross can be computed from the product of (Ss X Ss)(Aa X Aa) or

$$\frac{\begin{array}{c}1/4\ SS + 2/4\ Ss + 1/4\ ss\\ +\ 1/4\ AA + 2/4\ Aa + 1/4\ aa\end{array}}{}$$

1/16 SSAA + 2/16 SsAA + 2/16 SSAa + 4/16 SsAa + 1/16 SSaa + 2/16 Ssaa + 1/16 ssAA + 2/16 ssAa + 1/16 ssaa.

Each genotype was identified in the expected frequency following an appropriate test cross.

Mendel's second law or the law of independent assortment may be simply stated as follows: so long as allelic gene pairs are linked to separate chromosomes, they are assorted among the gametes each pair independent of the others.

5. Mendelian Genetics of Man

Mendelian genetic analyses in man are worthwhile for two reasons: First, such analyses are intrinsically valuable because they reveal how a particular trait is inherited. Second, knowledge of the mode of inheritance makes it possible to predict how often a particular trait will appear among the progeny of a specific marriage. This is especially important when harmful traits are involved. Since the number of progeny of a marriage is limited, the method of Mendelian analysis used depends on how frequently a specific trait occurs in a given population. In general, three situations

NOTES

are found: (1) The trait is rare and limited to one or only a few kindreds; (2) The trait is common and present in a number of kindreds; (3) The trait is universal and all individuals in a population can be classified as either having it or not having it. Examples of each situation follow.

Genetic conventions for kindred study: For traits of rare occurrence, the rules of transmission can often be deduced from the analysis of kindreds or family pedigrees. In the presentation of such pedigrees, a number of conventions and symbols have been adopted. The study of a particular trait often begins with a specific affected individual, the proband or propositus identified with an arrow (→) in each pedigree diagram. Males are designated by □ or ♂ and females by 0 or ♀. Individuals exhibiting the trait in question are indicated by a filled symbol ● or ♀, ■ or ♂. For any marriage males are commonly listed first and connected to their mate by a horizontal line □—0 or ♂—♀. Consanguineous marriages are indicated by a double horizontal line □═0. The progeny of each marriage are linked to the parents by a vertical line and chronological birth order is designated by Arabic numbers; each generation by a Roman numeral.

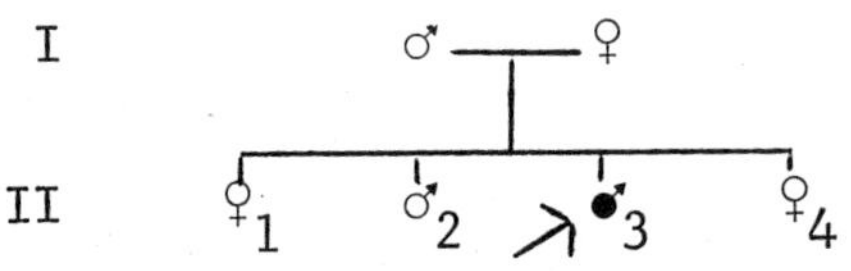

Individuals of unknown sex are designated by a diamond ◇ and stillbirths and abortions by a small, filled circle •. Monozygotic twins are designated by 0—0 or □—□ and dizygotic twins by 0 □ or □ □ or 0 0. More than one child of each sex is designated by [3] or (4).

Rare dominant, autosomal inheritance in man: A gene is defined as dominant when the phenotype of homozygotes and heterozygotes is indistinguishable. Where the trait is conditioned by a rare dominant gene, each affected individual is assumed to be heterozygous. The rarity of the trait precludes marriages between affected individuals, except in the unusual circumstance of a consanguineous marriage. Therefore, in general, all marriages will be of the type Aa X aa, where A is the dominant and a the recessive allele. The Mendelian expectations from such marriages are:

Figure 1. Seven Generations of Brachydactyly
(Unaffected Spouses Omitted)
(after Farabee, 1905)

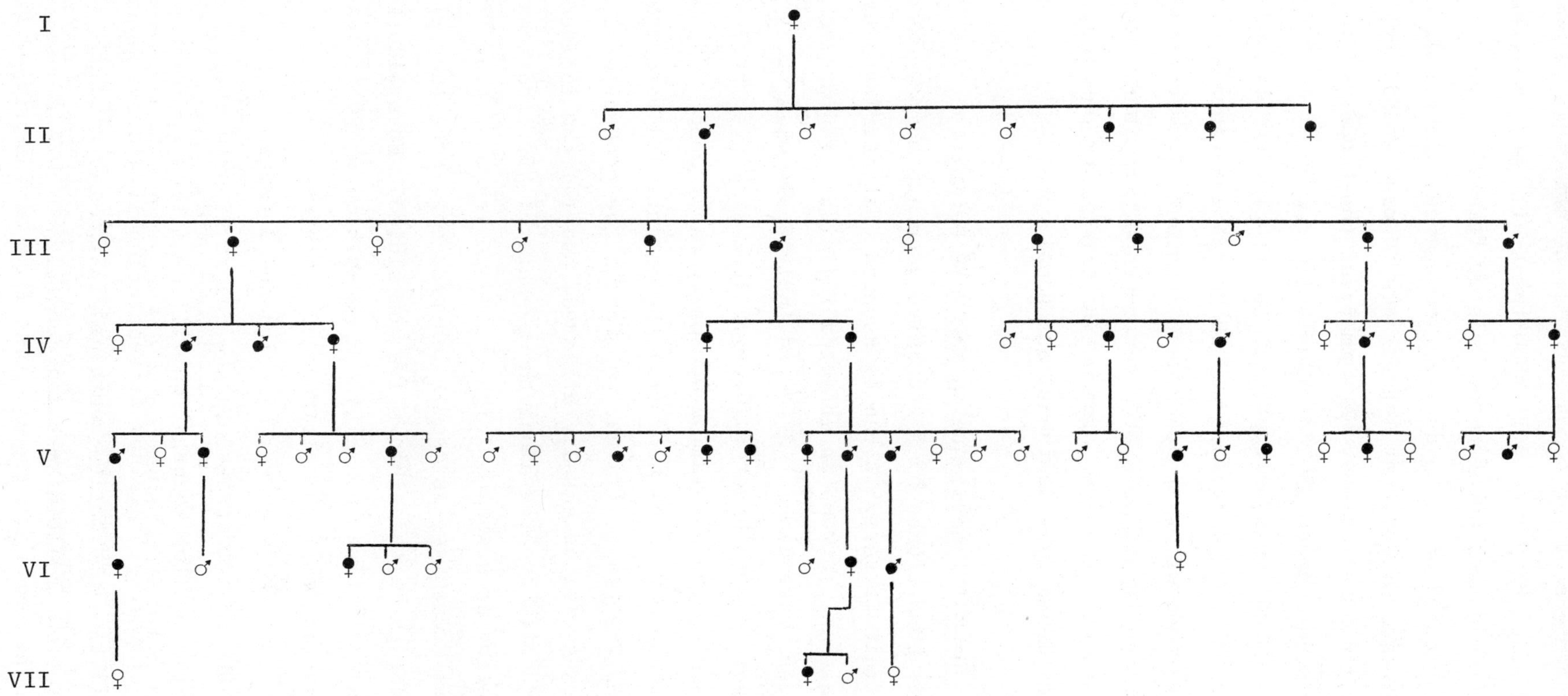

●♂ = brachydactylous

♀ = normal

(1) the trait will be transmitted directly from affected person to affected person, (2) equal numbers of the two sexes will be affected, (3) ca. one-half the progeny of an affected parent will show the trait, and (4) transmission of the trait occurs from either an affected male or female. Figure 1 is an unusually voluminous kindred of brachydactyly (= shortened fingers), apparently the first demonstration of dominant inheritance in man, and illustrates quite clearly the Mendelian expectations. A summary of the kindred data shows that (1) brachydactyly occurs among 39 individuals (14 ♂♂, 25 ♀♀); 43 individuals (24 ♂♂, 19 ♀♀) are normal in phenotype; (2) the progeny of brachydactyly females consist of 22 brachydactylous and 29 normal individuals; those of brachydactyly males consist of 17 brachydactylous and 12 normal individuals; (3) in this kindred the trait persisted for seven generations. Taken together, these facts mean that the best explanation for the brachydactyly phenotype is that it is conditioned by a dominant, autosomal gene.

Rare recessive, autosomal inheritance in man: By definition, individuals manifesting a trait conditioned by a recessive gene are homozygous for this gene. Since the trait is rare, it can be assumed that marriages between two affected individuals and between an affected individual and a heterozygote will rarely, if ever, take place. Affected individuals will occur among the progeny of marriages between heterozygotes or *Rr* X *Rr* since these are the most frequent individuals in the population who carry the *r* gene. The Mendelian expectations for these marriages are: (1) the parents and more remote relatives will not show the trait, (2) ca. one-fourth the progeny will show the trait, (3) both sexes will be equally affected, and (4) the incidence of consanguinity is frequently greater among the parents of affected individuals than in the population as a whole. Figure 2, a pedigree of albinism in Hopi Indians illustrates recessive, autosomal inheritance. Note that with but one exception the forebears of the albinos are non-albino. Note also that in this kindred the progeny of matings non-albino X non-albino where at least one child is albino consist of 11 non-albinos and 5 albinos. And note finally that albinos occurred in both sexes: 4 males and 3 females affected. The role of consanguinity is illustrated in the second kindred of albinism (figure 3). While two albino females (generation III) stem from apparently non-consanguineous marriages, all albinos of generation IV are the issue of consanguinity. All the information contained in the two kindreds is compatible with the interpretation of albinism being conditioned by a recessive, autosomal gene.

Figure 2. Pedigree of Albinism
(from Woolf and Dukepoo, 1969)

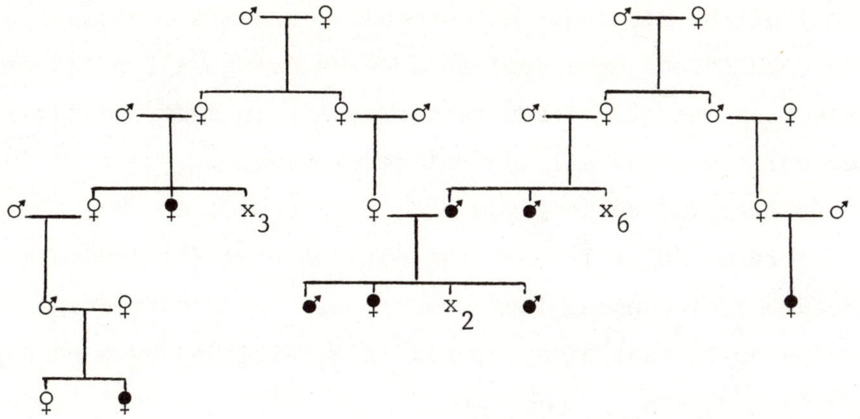

♂♀ = normal
●♂●♀ = albino
x = sex unknown

Figure 3. Pedigree of albinism with consanguinity
(from Pearson, Nettleship and Usher, 1911-13)

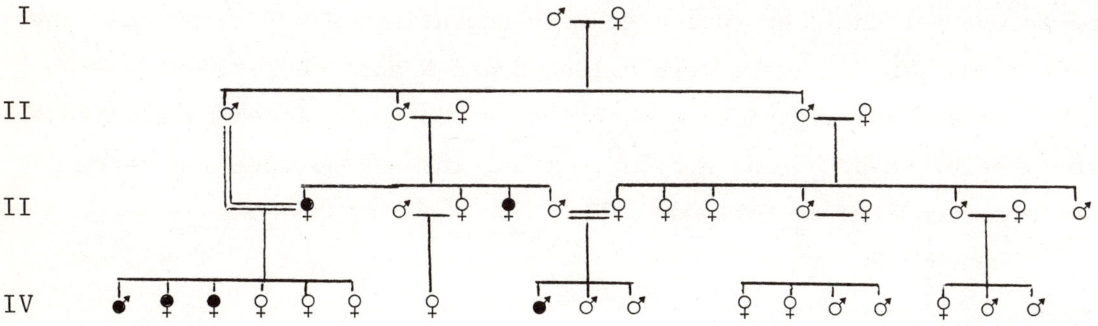

It should be emphasized that great care must be exercised in reaching conclusions from kindred data. Quite erroneous conclusions can be reached if the kindreds are incomplete. Just such a situation is illustrated by the accompanying kindred (figure 4) involving the trait, alkaptonuria. Considering only part (a) of the kindred, the conclusion appears inescapable that alkaptonuria is conditioned by a dominant, autosomal gene for transmission occurs regularly from one generation to the next. In fact, part (a) is an incomplete kindred, i.e., not all the related individuals were included. In part (b) the complete kindred is given and it is now clear that the conclusion based on part (a) requires revision. The data of part (b) indicate that a rather high incidence of consanguineous marriages has occurred and a more reasonable conclusion is that alkaptonuria is conditioned by a recessive, autosomal gene.

The relativity of dominance: The foregoing considerations of dominant and recessive alleles implies that heterozygous individuals are phenotypically indistinguishable from individuals homozygous for the dominant allele. Where an unequivocal distinction can be made, i.e., where each of the three genotypes can be identified in association with a distinctive phenotype, the alleles are said to be codominant or incompletely dominant. An interesting case in point is the M-N blood group system in man. By injecting red blood cells of different human beings into rabbits, two antisera (antibodies) were prepared, each specific for a particular antigen on the red blood cell surface. With these antisera, people were assigned to one of three classes according to whether their erythrocytes were agglutinated by one or the other or both antisera. The antigens were called M and N and the antisera anti-M and anti-N. It was postulated that each antigen is conditioned by a different, codominant allele: M by $\underline{a}^M$ and N by $\underline{a}^N$. Thus, three genotypes and three phenotypes are predicted as follows:

Phenotype	Response to anti-M	Response to anti-N	Genotype
M	+	-	$a^M a^M$
MN	+	+	$a^M a^N$
N	-	+	$a^N a^N$

Figure 4.

Imcomplete (a) and Complete (b) pedigrees of alkaptonuria
and the mode of inheritance of alkaptonuria
(from Khachadurian and Feisal, 1958)

(a)

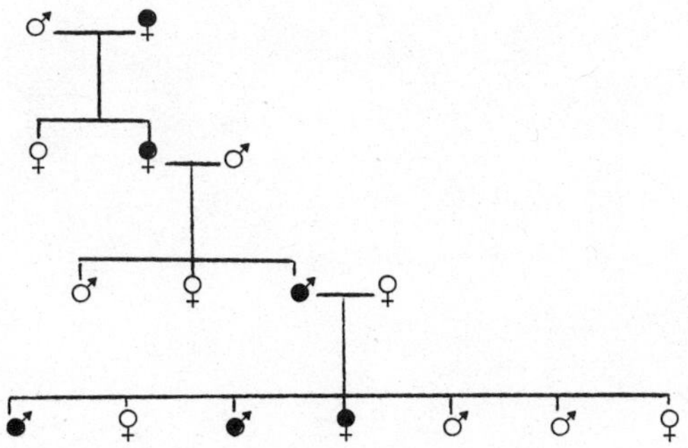

♀ = alkaptonuric

♂ = normal

(b)

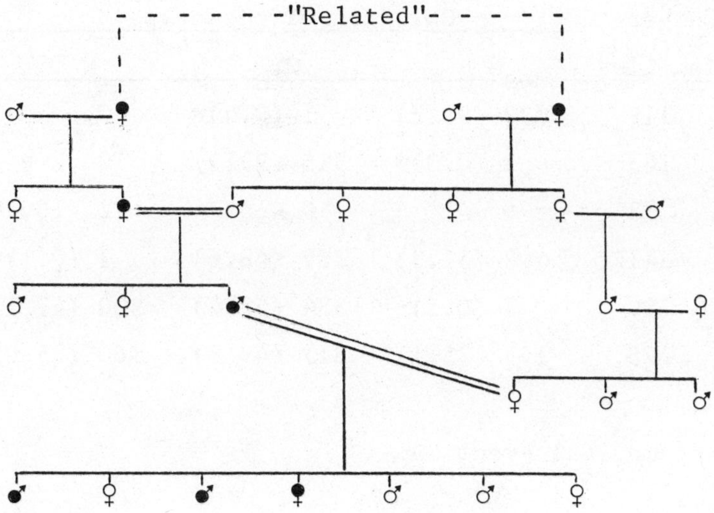

This hypothesis predicts the following distribution of children among the progeny of the several possible matings.

Parental	Frequency of children of phenotypes		
Phenotypes	M	MN	N
M X M	1		
M X N		1	
N X N			1
MN X M	½	½	
MN X N		½	½
MN X MN	¼	½	¼

Detailed family studies are in excellent agreement with prediction and confirm the explanation of codominant alleles. Representative data are included in the table which follows.

Distribution of M-N blood groups in 1160 families

Parental	Number	Children of phenotype			
Phenotypes	families	M	MN	N	Total
M X M	119	272 (99.6)	1 (0.4)*	0	273
M X N	143	1 (0.3)*	315 (99.7)	0	316
N X N	33	0	0	72 (100.0)	72
MN X M	341	408 (51.3)	387 (48.6)	1 (0.1)*	796
MN X N	250	3 (0.5)*	334 (52.4)	300 (47.1)	637
MN X MN	275	163 (25.1)	317 (49.5)	160 (25.0)	640

*probably extramarital event.

The M-N blood group system is a model system of codominance. More often the phenotypic distinction between homozygote and heterozygote is much subtler, often seemingly trivial. Such is the case for sickle cell anemia, an anemia usually fatal in early childhood. In kindreds of sickle cell anemia, two types of normal individuals are identified on the basis of their erythrocyte morphology when the oxygen tension is reduced by sealing a drop

Figure 5. Two kindreds with sickle cell anemia (after Neel, 1951)

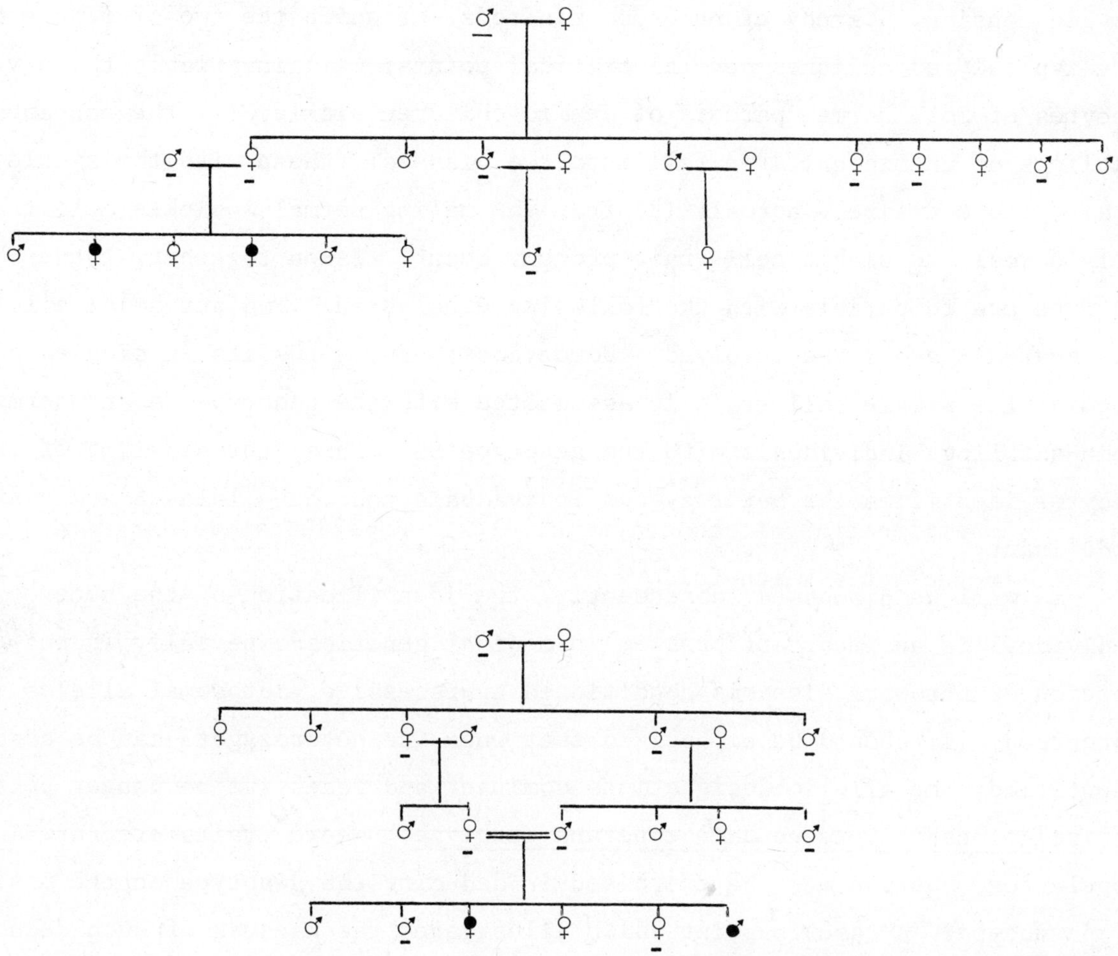

♀♂ = sickle cell trait (symbols underlined)

●♀ ●♂ (filled symbols) = sickle cell anemia

of blood on a microscope slide for a few hours. The erythrocytes of some individuals remain unchanged; those of other people sickle, i.e., change their morphology from a disc form to a sickle form (sickle cell trait). The erythrocytes of anemic individuals also undergo a striking sickling with reduced oxygen tension. A study of numerous kindreds, of which the two of figure 5 are typical, establishes several critical points: (1) invariably the erythrocytes of both normal parents of anemic children sickle; (2) the non-anemic siblings of anemic children fall into two classes: those with the sickle cell trait, those entirely normal; (3) from the mating normal X sickle cell trait only normal and sickle cell trait progeny occur. Taken together, these observations are compatible with the following conclusion. Two autosomal alleles designated S and s are involved. Homozygosity for s results in sickle cell anemia; the sickle cell trait is associated with the genotype Ss and normal (non-sickling) individuals with the genotype SS. Thus, the sickling of erythrocytes identifies the heterozygous individuals and the alleles S and s are codominant.

As will be discussed subsequently, the identification of the heterozygous individual is an important problem in medical genetics especially where prevention of untoward diseases conditioned by recessive, autosomal alleles is concerned. It should be emphasized that when the heterozygote can be accurately identified, the allelic designations dominant and recessive no longer pertain.

The relationship between phenotype and genotype: Where traits are rare in the population, caution must be exercised in deducing the genotype on the basis of phenotype. A case in point which illustrates the hazards of such deductions is illustrated in figure 6, a kindred of albinism. As discussed earlier (cf. figures 2 and 3) albinism is best explained as being conditioned by a relatively rare, autosomal recessive gene. On this basis the albino individuals which segregate in generation IV should be of identical genotype viz. homozygous for the recessive albinism gene. Serious doubt as to the correctness of this explanation stems from the fact that the marriage of two unrelated albinos (IV-3 X IV-4) resulted in a normally pigmented daughter. Assuming this daughter is the natural offspring of this marriage--and this was convincingly established--how can the phenotype and genotype of the daughter be explained. The most reasonable explanation is that the albinism phenotype can be conditioned by two different, unrelated recessive autosomal genes. In this kindred both types of albinism genes segregate. Thus individuals IV-4, IV-10 and IV-11 are homozygous for the gene a, but individual IV-3 is homozygous for the gene c. Assuming a and c are not alleles, and assuming each gene is relatively

Figure 6. Non-allelic albinism = Complementary genes

(Witkop et al. Amer. J. Human Genetics 22:55, 1970)

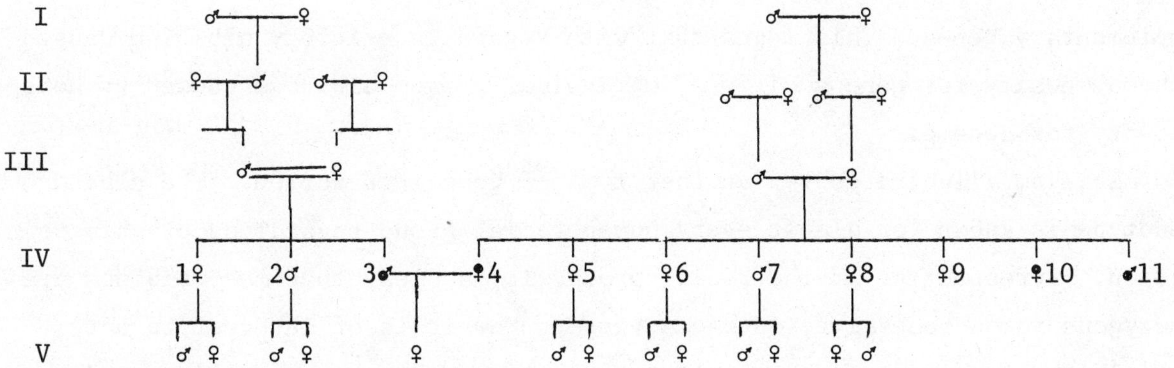

● = albino

♀ = normal

rare, the genotypes of the albino individuals are deduced as follows. The complete genotype of IV-4, IV-10 and IV-11 is aa;CC; that of IV-3 is AA;cc. Thus all the progeny of the marriage IV-3 (AA;cc) X IV-4 (aa;CC) will be Aa;Cc or normal pigmentation in phenotype. The genes a and c are said to be complementary genes. This means that with regard to etiology albinism caused by homozygosity for gene a differs intrinsically from albinism caused by homozygosity for gene c.

Albinism illustrates yet another type of gene interaction. The albino phenotype is known for almost every human racial group regardless of skin coloration. Irrespective of the racial group, it is clear that most albinos are homozygous for a recessive, autosomal gene. Now it is of interest to note that albinos occurring among socalled Caucasians, Negroes, Mongolians and American Indians are on the basis of skin color indistinguishable despite the striking differences in skin coloration of these racial groups. The albino gene is said to be epistatic to all other pigmentation genes, that is, when homozygous it supercedes or "dominates" all other pigmentation genes and it alone effectively determines the phenotype.

Gene, environment and phenotype: In the preceding discussions of Mendelian genotypes and phenotypes two tacit assumptions were made. First, the determination of an individual's phenotype from zygote to parturition is the product of the biological reaction: genotype + environment → phenotype. Second, the role of the environment in this reaction scheme is relatively trivial, i.e., in spite of fluctuation in the environment, each genotype results in a characteristic phenotype. Or stated another way, to understand and define Mendelian segregation, it was necessary to deliberately choose allele combinations effectively insensitive to environmental changes and where the genotype is all-important in the realization of the phenotype. It took the genius of Gregor Mendel to recognize this fact, for without this recognition the Mendelian laws of genetic transmission could hardly have been deduced. Nonetheless, it is recognized that the environment often does have a critical part to play in phenotype differentiation such that for a given genotype, depending upon the environmental effects, the same phenotype need not be manifest. Consequently, the term penetrance was introduced to denote the relationship between genotype and phenotype. Where penetrance is complete, 100 percent of the individuals of a specified genotype exhibit the same phenotype. Where penetrance is incomplete, some percentage (less than 100%) of individuals of a given genotype will manifest the specified phenotype (usually abnormal) while others will be normal. Polydactyly in man

Figure 6a.

A kindred of polydactyly with incomplete penetrance

(after Neel and Schull, 1952)

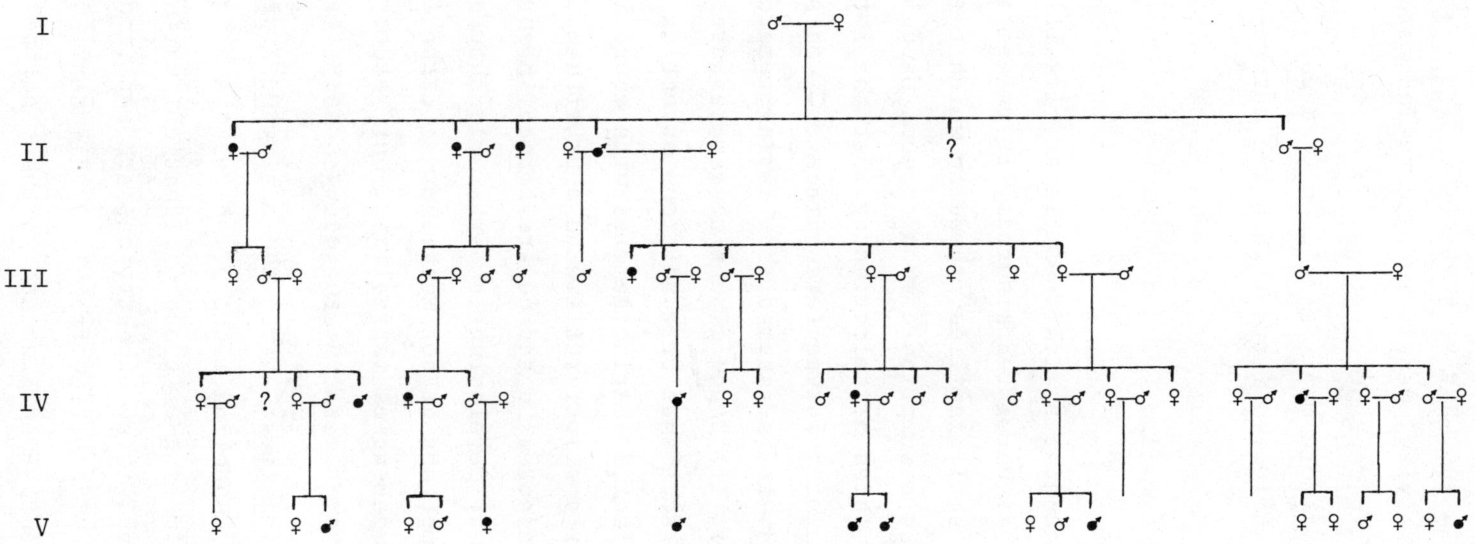

● = polydactyly

♀ = normal

represents a trait whose penetrance is incomplete. The kindred given in figure 6a illustrates the incomplete penetrance associated with polydactyly. Note that while some polydactylous individuals have a polydactylous parent, others do not. Since the trait is rare and since there is little evidence of consanguinity, the simplest explanation for polydactyly is that it is conditioned by a dominant gene with incomplete penetrance.

Yet another term, expressivity, was introduced to indicate the degree of phenotypic expression of a penetrant genotype. The expressivity may range from slight to severe and is also a manifestation of the environmental influences. Both terms will be considered further when the nature-nurture problem is discussed.

6. Genetic Ratios in Small Families

Before considering the genetic analysis of traits classified as common and thus present in a number of unrelated kindreds, it is necessary to examine Mendelian segregation in small families, those in which the number of progeny varies from 1-10. One assumption common to most Mendelian analyses is that chance abounds and no bias or selection against the formation of any gamete or zygote irrespective of genotype takes place. Given a heterozygous individual of the genotype Dd, it is assumed that during gametogenesis the following conditions prevail: (1) D and d bearing gametes are equally frequent, (2) any developmental accidents or conditions affecting the formation of D bearing gametes will affect d bearing gametes with equal likelihood, (3) spermatozoa carrying D or d will fertilize an ovum with equal probability, (4) from the cross Dd X dd, zygotes Dd and dd are formed in equal frequency and have equal probabilities of developing to birth. It should be emphasized that the aforementioned assumptions do not mean that an exact 1:1 relationship in gametic or zygotic frequencies is realized, but rather than whatever deviations from 1:1 occur, they are random and occur solely through chance.

Segregation ratios in small families: Given the mating Dd X dd where D and d are autosomal alleles, the Mendelian expectation for the progeny is $\frac{1}{2}$Dd and $\frac{1}{2}$dd. Where from any one mating or group of identical matings large numbers of progeny (>100) are obtained, these expectations are realized. What is the situation for families whose size varies from 1-10 children? Note that each child represents an independent genetic event and the genotype of one child in no way influences the genotype of any other child. For families with one child, the expectations are obvious: in half of such families the child will be Dd, in half the child will be dd. For sibships of two children, the

NOTES

expectations are as follows. The probability that the first child will be Dd is one-half, or dd also one-half. For the second child exactly the same probabilities apply. These events being independent, the chance that both children will be Dd is $\frac{1}{2} \times \frac{1}{2}$ or $\frac{1}{4}$, that one child will be Dd, the other dd is $2(\frac{1}{2} \times \frac{1}{2})$ or $\frac{1}{2}$, and both children will be dd is $\frac{1}{2} \times \frac{1}{2}$ or $\frac{1}{4}$. These probabilities are the expansion of the binomial $(\frac{1}{2}\underline{Dd} + \frac{1}{2}\underline{dd})^2$. If all sibships of two are considered together, the progeny comport to the Mendelian expectation of 1Dd:1dd.

Consider next sibships of size 4, derived from the same mating Dd X dd. The conception of each child being an independent genetic event, for such sibships five different distributions of Dd and dd children are expected in the following frequencies:

Distribution of Children	Frequency
4 Dd	$(\frac{1}{2} \times \frac{1}{2} \times \frac{1}{2} \times \frac{1}{2}) = 1/16$
3 Dd + 1 dd	$4(\frac{1}{2} \times \frac{1}{2} \times \frac{1}{2} \times \frac{1}{2}) = 4/16$
2 Dd + 2 dd	$6(\frac{1}{2} \times \frac{1}{2} \times \frac{1}{2} \times \frac{1}{2}) = 6/16$
1 Dd + 3 dd	$4(\frac{1}{2} \times \frac{1}{2} \times \frac{1}{2} \times \frac{1}{2}) = 4/16$
4 dd	$(\frac{1}{2} \times \frac{1}{2} \times \frac{1}{2} \times \frac{1}{2}) = 1/16$

Note that these probabilities are the expansion of the binomial $(\frac{1}{2}\underline{Dd} + \frac{1}{2}\underline{dd})^4$. Note further that of these sibships, only 6/16 fulfill the ideal Mendelian expectation of 1Dd:1dd. However, when all sibships of 4 are considered collectively, the progeny are distributed in the frequency 32Dd:32dd or the Mendelian expectation of 1:1. As the sibship size increases, the number of sibships equalling or approximating the 1:1 expectation also increases. It is now possible to generalize as follows. For the mating Dd X dd the probability of realizing each of the different combinations of Dd and dd progeny in a sibship of size $\underline{n}$ is described by expanding the binomial $(\frac{1}{2}\underline{Dd} + \frac{1}{2}\underline{dd})^n$. The frequency of any one combination of Dd and dd children can be calculated from the relationship $(\frac{n!}{a!b!})(\frac{1}{2}\underline{Dd})^a(\frac{1}{2}\underline{dd})^b$ where $\underline{n}$ equals the total number of children in the sibship, $\underline{a}$ equals the number of Dd children and $\underline{b}$ equals the number of dd children.

Consider now the mating Dd X Dd. What are the expected distributions of genotypes for sibships of different sizes? The Mendelian expectation for the progeny of this mating is $\frac{3}{4}\underline{D-}$ and $\frac{1}{4}\underline{dd}$ where D- represents the sum of DD and Dd individuals. For sibships of size 2, 3 and 4 the distribution and frequency of the two kinds of children are given in the following table:

NOTES

Sibship size	Distribution of Children	Frequency
2	2 D-	$(\frac{3}{4} \times \frac{3}{4}) = 9/16$
	1 D- and 1 dd	$2(\frac{3}{4} \times \frac{1}{4}) = 6/16$
	2 dd	$(\frac{1}{4} \times \frac{1}{4}) = 1/16$
3	3 D-	$(\frac{3}{4} \times \frac{3}{4} \times \frac{3}{4}) = 27/64$
	2 D- and 1 dd	$3(\frac{3}{4} \times \frac{3}{4} \times \frac{1}{4}) = 27/64$
	1 D- and 2 dd	$3(\frac{3}{4} \times \frac{1}{4} \times \frac{1}{4}) = 9/64$
	3 dd	$(\frac{1}{4} \times \frac{1}{4} \times \frac{1}{4}) = 1/64$
4	4 D-	$(\frac{3}{4} \times \frac{3}{4} \times \frac{3}{4} \times \frac{3}{4}) = 81/256$
	3 D- and 1 dd	$4(\frac{3}{4} \times \frac{3}{4} \times \frac{3}{4} \times \frac{1}{4}) = 108/256$
	2 D- and 2 dd	$6(\frac{3}{4} \times \frac{3}{4} \times \frac{1}{4} \times \frac{1}{4}) = 54/256$
	1 D- and 3 dd	$4(\frac{3}{4} \times \frac{1}{4} \times \frac{1}{4} \times \frac{1}{4}) = 12/256$
	4 dd	$(\frac{1}{4} \times \frac{1}{4} \times \frac{1}{4} \times \frac{1}{4}) = 1/256$

Note that in none of the sibships of sizes 2 and 3 is the Mendelian expectation of $\frac{3}{4}$D- and $\frac{1}{4}$dd fulfilled. However, taken altogether, the individuals in each sibship comport to a 3D-:1dd frequency. For the sibships of 4 children, 108/256 fulfill the Mendelian 3D-:1dd. Similarly taking all the individuals together, the expectation of 3D-:1dd is realized. Again, as the sibship size increases, more and more families fulfill Mendelian expectation. For the mating Dd X Dd the probability of realizing each of the different combinations of D- and dd progeny in a sibship of size n is described by expanding the binomial $(\frac{3}{4}D- + \frac{1}{4}dd)^n$. The frequency of any one combination of Dd and dd children can be calculated from the relationship $(\frac{n!}{a!b!})(\frac{3}{4}D-)^a(\frac{1}{4}dd)^b$ where n equals the sibship size, a the number of D- children and b the number of dd children.

The foregoing information provides a basis for defining the mode of inheritance of a common trait. It is clear that sibship size, per se, does not preclude the occurrence or detection of Mendelian segregation. Given a sufficient number of sibships of any size, it should be possible to determine whether or not any common trait conforms to a Mendelian mode of inheritance. By and large two different genetic situations can be anticipated: the trait is conditioned by an autosomal dominant or the trait is conditioned by an autosomal recessive.

NOTES

For a common trait conditioned by a dominant autosomal gene, essentially all individuals in the population manifesting the trait will be heterozygous and, barring consanguinity, will marry normal (homozygous recessive) individuals. This, because the trait although common, is relatively infrequent in the population. Assuming complete penetrance, if a population is thoroughly censused, all matings of the type Dd X dd can be identified on the basis of phenotype. In other words all appropriate matings are completely ascertainable. If, in fact, the common trait is conditioned by an autosomal dominant, then a 1:1 segregation should be realized among the progeny of the ascertained matings.

The inheritance of the socalled nail-patella syndrome serves to illustrate complete ascertainment. Kindred data impute a dominant, autosomal gene. A series of matings of the type nail-patella syndrome X normal were ascertained and progeny from each recorded. The results are presented in Table 1a and are unequivocal. Among 157 different sibships with a total of 556 progeny, 268 were of the syndrome phenotype and 288 normal. Statistically, the results are not different from 1:1 and it is reasonable to conclude that the syndrome is conditioned by a dominant autosomal gene.

Alkaptonuria is an example of a common trait conditioned by a recessive, autosomal gene. Affected individuals excrete in their urine a compound called homogentisic acid. On exposure to air homogentisic acid rapidly oxidizes to alkapton, a black pigment. Thus, among alkaptonurics, almost immediately after urination their urine turns black, a frightening though harmless event. (The metabolic defect associated with alkaptonuria will be considered in detail later.) Kindred studies show that most alkaptonurics are the progeny of normal parents and that the incidence of alkaptonuria increases with consanguinity. As a first approximation, these facts suggest that segregation of an autosomal, recessive gene accounts for the occurrence of alkaptonuria. A more precise test of this explanation involves censusing as many kindreds as possible for the occurrence of alkaptonurics and determining statistically whether the observed incidence is compatible with the Mendelian hypothesis.

If alkaptonuria is conditioned by an autosomal, recessive allele a, it can be assumed in the absence of consanguinity that most alkaptonurics (aa) stem from matings of Aa X Aa. As demonstrated above, if all possible matings of this type are identified (ascertained), the expected phenotypes among the progeny will approximate 3/4 normal, 1/4 alkaptonuria. How are such matings ascertained? In practice, a mating of the type Aa X Aa will be ascertained by the occurrence of one or more alkaptonuric progeny. This means that those

Table 1a.

Nail-patella Syndrome--Parents = Syndrome X Normal

Number sibships	Number progeny	Phenotype Progeny Syndrome	Normal
78	264	141	123
15	53	24	29
8	36	11	25
17	79	35	44
33	99	42	57
2	11	7	4
4	14	8	6
157	556	268	288

matings of Aa X Aa in which by chance no alkaptonuric progeny occur will not be ascertained unless some method of detection other than the alkaptonuria phenotype is employed. What is the Mendelian segregation expected when these matings are not included in the census? As an illustration, consider sibships with 4 offspring. The ideal Mendelian expectation is:

Distribution of		Number of Progeny	
Progeny Phenotype	Frequency	Normal	Alkaptonuric
4 normal	81/256	324	0
3 normal, 1 alkaptonuric	108/256	324	108
2 normal, 2 alkaptonuric	54/256	108	108
1 normal, 3 alkaptonuric	12/256	12	36
4 alkaptonuric	1/256	0	4
	TOTAL	768	256

In the ideal situation where all sibships of 4 are ascertained, the progeny will consist of 768 normal (A-): 256 alkaptonurics (aa). This reduces to 3:1, the classical Mendelian ratio. If all sibships of 4 lacking alkaptonuric progeny are not ascertained, the Mendelian expectation becomes (768-324) normal (A-): 256 alkaptonurics (aa) or 444:256, a significant deviation from 3:1. This means that which matings of the type Aa X Aa are ascertained has an important bearing on the phenotypic expectations. When selection of Aa X Aa matings depends solely upon the ascertainment of alkaptonuric progeny, the data will, by necessity, be incomplete because those sibships lacking alkaptonurics will be overlooked. With incomplete selection, three possible methods of identifying appropriate matings occur:

(1) truncate selection - every individual is independently ascertained as though the population were exhaustively examined. A family with a large number of children is no more likely to be included than is a family with one or few affected children. Here the distribution of affected children is approximately binomial except for the omission of the matings with no affected children which are truncated.

(2) single selection - the method of ascertainment is such that no family is included more than once. This might occur where all the children of one school grade are surveyed. Here the probability of a mating being ascertained is proportional to the number of affected progeny.

(3) multiple selection - this is a combination of (1) and (2).

NOTES

Testing the recessive hypothesis: For illustrative purposes only incomplete ascertainment with truncate selection will be considered. Since truncate selection excludes those Aa X Aa matings where no aa progeny occur, the sibships included will be biased and the frequency of aa sibs will be increased proportionally. Referring again to sibships of 4, when all matings are included the Mendelian segregation is 768 A-:256 aa; with truncate selection the segregation ratio becomes 444 A-:256 aa. Can the bias imposed by truncate selection be rationalized and a precise test of the recessive mode of inheritance made? Two methods will be discussed which rationalize the bias but in quite different ways.

Weinberg sib or proband method: The rationale of this method is to correct for the excess aa individuals imposed by incomplete ascertainment. For the cross Aa X Aa the probability that any one offspring will be normal (AA or Aa) is 3/4 or alkaptonuric (aa) is 1/4. Since each offspring represents an independent genetic event, the probability that a sib of an alkaptonuric will be normal is also 3/4 or alkaptonuric is also 1/4. Hence if alkaptonuria is conditioned by an autosomal recessive gene, the ratio of normals:alkaptonurics among the sibs of an alkaptonuric should approximate the Mendelian expectation of 3:1 provided the proband alkaptonuric is not included in the sibship total. Inclusion of the proband biases the ratio in favor of alkaptonurics. Therefore the proband is excluded to correct for this bias. Furthermore each sibship will be included in the evaluation as often as it can be independently ascertained, or equal to the number of probands (alkaptonurics) in the sibship. If a sibship includes only one alkaptonuric, it can be ascertained only through this one proband. Thus the sibship will be "scored" only once after first excluding the proband. If a sibship includes two alkaptonurics, it has two probands and can be ascertained twice, once through one proband, once through the other. This sibship will be "scored twice, i. e., once for each proband after first excluding the proband. If a sibship includes as many as five alkaptonurics, it has five probands and will be "scored" five times, once for each proband. The proportion of alkaptonurics among the "scored" progeny should approximate 1:3 if alkaptonuria is conditioned by an autosomal recessive gene.

The working of the method may be demonstrated by using families of four derived from the cross Aa X Aa. A "Sibship score" which represents the corrected number of normal and alkaptonuric sibs after excluding the proband is computed for each sibship. These are given in the following table in which the ideal situation is represented:

NOTES

Family of 4:		Sibship score	
Sibs with alkaptonuria	Frequency	Normal	Alkaptonuria
0	81/256		
1	108/256	324	0
2	54/256	108 + 108 = 216	54 + 54 = 108
3	12/256	12 + 12 + 12 = 36	24 + 24 + 24 = 72
4	1/256		3 + 3 + 3 + 3 = 12
		TOTAL 576	192

The families with 0 alkaptonurics were omitted from the sibship score since they cannot be ascertained. A family with 1 alkaptonuric is ascertained through this sib and the "score" includes no affected sibs since the proband is excluded. Families with two alkaptonurics have two probands and hence are "scored" twice, once for each proband. Similarly families with 3 and 4 alkaptonurics have 3 and 4 probands and thus are "scored" 3 and 4 times. The sum of the sibship score is 576 normal:192 alkaptonuria or 3:1, the Mendelian segregation.

In practice, the same procedure is followed and the proportion of alkaptonurics in segregating families is determined after correcting for the probands. If the computed proportion of alkaptonurics approximates the $\frac{1}{4}$ expected for a simple autosomal, recessive, it is so concluded. In the following table a number of families of 5, 6 and 7 children segregating for alkaptonuria are given:

Number of sibs with alkaptonuria	Number of Children/Family 5	6	7
1	7	4	4
2	6	6	4
3	4	3	5
4	1	1	1

Assuming complete ascertainment, consider first families of 5 children. The data may be "scored" in the following manner as noted in the table below:

NOTES

Families of 5 children

Number with Alkaptonuria	Number families	Sibship Scores Normal	Sibship Scores Alkaptonuria
1	7	28 (4 X 7)	0
2	6	36 (3 X 6 X 2)	12 (1 X 6 X 2)
3	4	24 (2 X 4 X 3)	24 (2 X 4 X 3)
4	1	4 (1 X 1 X 4)	12 (3 X 1 X 4)
	TOTAL	92	48

For all families of 5, y, the proportion of alkaptonurics computed from the sibship scores equals $\frac{48}{(92 + 48)}$ or 0.342. By employing the same procedure y for all families of 6 is 0.290, and y for all families of 7 is 0.275. The mean y equals 0.299. This can be tested statistically and shown not to differ significantly from the expected y = 0.25. From these results it is concluded that alkaptonuria segregates as an autosomal, recessive gene.

A priori or direct method of Hogben-Lenz: The rationale of this method is simple. As demonstrated earlier, truncate selection biases the Mendelian expectation for homozygous aa derived from the cross Aa X Aa so that the fraction of aa individuals will depend on the sibship's size (s) and the proportion of sibships not ascertained (x). Assuming truncate selection of the appropriate families, the proportion of sibships which contain no aa individuals, i.e. those which cannot be ascertained (x) is $(3/4)^s$. The number of sibships containing at least one aa individual i.e. those which can be ascertained (y) is $1-(3/4)^s$. Thus for any sibship of size s the expected proportion (z) of aa individuals equals the quotient $\frac{1/4}{y}$. This quotient, being a function of the size of s, is different for every value of s. Where s is small (e.g. 1, 2 or 3) the deviation from the Mendelian expectation of 1/4 is large (cf. table 1); where s is large (e.g. 12) the deviation from 1/4 is small (cf. table 1). In table 1, z and sz, the average number of homozygous recessive individuals for each sibship, are given. In practice, the number of alkaptonurics in all families of each sibship size s is determined. The observed frequency of alkaptonurics is compared with the expected (calculated) frequency assuming ascertainment is incomplete. If alkaptonuria is conditioned by a recessive, autosomal allele, the observed and expected frequencies should not differ significantly. Just such a comparison is given in table 2. The difference between the observed and expected frequencies of alkaptonuria is not statistically significant. Thus, the transmission of alkaptonuria is best explained by concluding the trait is conditioned by a recessive, autosomal gene.

Table 1. Testing the recessive mode of inheritance by the *a priori* (Hogben-Lenz) method

	Predicted values			
	Proportion of Sibships			
Sibship Size	Cannot be Ascertained	Can be Ascertained	Proportion Homozygous Recessives	Average Number Homozygous Recessives
s	$[x = (3/4)^s]$	$[y = 1-(3/4)^s]$	$z = \frac{1/4}{y}$	sz
1	3/4	1/4	1	1
2	$(3/4)^2 = 9/16$	7/16	4/7 = 0.5714	1.143
3	$(3/4)^3 = 27/64$	37/64	16/37 = 0.4324	1.297
4	$(3/4)^4 = 81/256$	175/256	64/175 = 0.3657	1.463
5	$(3/4)^5$		0.3278	1.639
6	$(3/4)^6$		0.3041	1.825
7	$(3/4)^7$		0.2885	2.020
.				
.				
.				
.				
.				
12	$(3/4)^{12}$		0.2582	3.099

7. Mendelism in Populations

An understanding of the genetic basis of a number of universal human traits, those for which all individuals can be classified, has been greatly facilitated by the application of the principles of population genetics. Before considering these principles, it is necessary to define two concepts: that of a population and that of a Mendelian population. A population can be simply defined as all the individuals occupying a specific geographical area. Alternative definitions of a Mendelian population are: (1) a community of interbreeding individuals, or (2) a group of individuals who share a common gene pool or freely exchange their genes, or (3) the local representation of that level of biological organization whose adjacent lower level is the individual and whose adjacent higher level is the species.

Among the factors which influence the genetic composition of a population, the most important are: (1) selection, (2) mutation, (3) the system of mating, (4) migration, and (5) genetic drift. The influence of each will be considered.

Systems of mating, gene frequency, genotypes and phenotypes: Before discussing the influence of the mating system on the genetic structure of populations, it is necessary to introduce the concept of gene or allele frequency. This concept will be illustrated by considering the inheritance of agouti (wild-type) and albino coat colors in the mouse. For purposes here agouti is conditioned by the dominant autosomal allele A and albino by the recessive allele a. Beginning with homozygous lines the following crosses were made with these results:

$$P \quad \underline{AA} \times \underline{aa}$$
$$F_1 \quad \text{all } \underline{Aa}$$
$$F_1 \times F_1 \rightarrow F_2 \quad 1/4\ \underline{AA} + 2/4\ \underline{Aa} + 1/4\ \underline{aa}$$
$$\text{or } F_2 \quad 3/4 \text{ agouti} + 1/4 \text{ albino.}$$

So far as the P generation is concerned, there are two alleles in equal frequency present. Thus, we may specify that in the P generation, the allele frequencies are: 0.5 A + 0.5 a = 1. The allele frequencies in the F_1 are unchanged for all the individuals are heterozygous - and heterozygotes by definition possess contrasting alleles in equal frequency. Among the F_2 the allele frequencies remain the same: 0.5 A and 0.5 a.

Table 2.

Alkaptonuria - Offspring of two normal parents

Family size s	n	Alkaptonuria Observed	Expected	Calculations of expected (=z.ns)
1	5	5	5	5 x 1
2	8	10	9.14	0.5714 x 16 = 9.14
3	5	8	6.48	0.4324 x 15 = 6.48
4	2	4	2.93	
5	3	4	4.92	
6	3	8	5.47	
7	2	5	4.04	
8	3	5	6.67	
9	1	4	2.43	
10	1	2	2.65	
11	3	7	8.61	
14	1	4	3.56	
Σ	37	66	61.90	Difference not significant statistically

Random mating (panmixia): Suppose all the F_2 individuals were bred at random. What is the expectation in the F_3? As spelled out in the crossing matrix below, the prediction is that the F_3 population will be exactly the same as the F_2, viz. 1/4 AA + 2/4 Aa + 1/4 aa.

		F_2 ♂♂ 1/4 AA	2/4 Aa	1/4 aa
	1/4 AA	1/16 AA	1/16 AA 1/16 Aa	1/16 Aa
F_2 ♀♀	2/4 Aa	1/16 AA 1/16 Aa	1/16 AA 2/16 Aa 1/16 aa	1/16 Aa 1/16 aa
	1/4 aa	1/16 Aa	1/16 Aa 1/16 aa	1/16 aa

$$F_3 = \begin{array}{l} AA = 1/16 + 1/16 + 1/16 + 1/16 = 4/16 \text{ or } 1/4 \\ Aa = 1/16 + 1/16 + 1/16 + 2/16 + 1/16 + 1/16 + 1/16 = 8/16 \text{ or } 2/4 \\ aa = 1/16 + 1/16 + 1/16 + 1/16 = 4/16 \text{ or } 1/4 \end{array}$$

Note that the allele frequencies are unchanged: 0.5 A + 0.5 a. Random breeding of the F_3 individuals gives an F_4 identical to the F_3 population in allele frequenies, genotypes and phenotypes. Thus, this randomly breeding population is said to be at equilibrium. The maintenance of the equilibrium in this idealized condition requires the fulfillment of a number of conditions. These include: (1) mutation does not occur, or if it does, the frequency of mutations of A → a is equal to the mutation frequency of a → A; (2) segregation is regular; (3) no selection of gametes occurs; (4) fertilization is random; (5) all zygotes and embryos are equally viable; (6) all adults to reproductive age are equally viable; (7) no differential sexual mortality occurs; (8) no differential fertility occurs; (9) no overlapping of generations occurs; (10) panmixia or random breeding of the population occurs; (11) the population approximates an infinite size; (12) no migration into or out of the population occurs.

Assortative (non-random) mating: In assortative matings individuals of a particular type breed together more often than expected from their

NOTES

frequencies in the population. This can be illustrated by returning to the F_2 generation of mice consisting of 1/4 AA + 2/4 Aa + 1/4 aa individuals. Suppose now matings occur only between individuals of identical genotype. What is the expectation for the F_3? Provided the aforementioned assumptions as regards mutation, selection and fertility obtain, it can be shown that the F_3 generation will consist of 3/8 AA + 2/8 Aa + 3/8 aa individuals. Note that the allele frequencies of 0.5 A + 0.5 a are unchanged. Thus the change in frequency of genotypes, and hence phenotypes, is a function of the mating system. The further positive assortative mating of the F_3 individuals results in a further halving of the Aa frequency giving an F_4 of 7/16 AA + 2/16 Aa + 7/16 aa.

Genetic equilibrium — the Hardy-Weinberg Law: Suppose a new population was started from parents consisting of 80% homozygous agouti and 20% albino mice. The agouti and albino mice include equal numbers of males and females. In terms of gene frequency the population consists of 0.8 A + 0.2 a = 1. If mating is wholly random, the following genotypic and phenotypic results are predicted for the next generation:

		sperm 0.8 A	sperm 0.2 a
ova	0.8 A	0.64 AA	0.16 Aa
ova	0.2 a	0.16 Aa	0.04 aa

genotypic frequency 0.64 AA + 0.32 Aa + 0.04 aa = 1

phenotypic frequency 0.96 agouti + 0.04 albino = 1

Note that the allele frequencies remain 0.8 A + 0.2 a = 1. Suppose the population of 0.64 AA + 0.32 Aa + 0.04 aa mice is allowed to breed at random. The genetic composition of the next generation will be:

NOTES

			Frequency of Progeny		
Mating	Frequency of Mating		AA	Aa	aa
AA x AA	0.64 x 0.64	= 0.4096	0.4096		
AA x Aa	2(0.64 x 0.32)	= 0.4096	0.2048	0.2048	
AA x aa	2(0.64 x 0.04)	= 0.0512		0.0512	
Aa x Aa	0.32 x 0.32	= 0.1024	0.0256	0.0512	0.0256
Aa x aa	2(0.32 x 0.04)	= 0.0256		0.0128	0.0128
aa x aa	0.04 x 0.04	= 0.0016			0.0016
		TOTAL	0.64	0.32	0.04

Note that the genotypic and phenotypic frequencies of the population are identical to those of the preceding generation. Subsequent random breeding of this population gives precisely the same genotypic and phenotypic frequencies. Thus, so long as the allele frequencies (gene pool) remain unchanged and breeding is panmictic, the genotypes (and hence phenotypes) of the population will be in genetic equilibrium. For the allele frequencies 0.8 A + 0.2 a, the equilibrium genotypes are: 0.64 AA + 0.32 Aa + 0.04 aa = 1. This generalization was independently described in 1908 by the English mathematician, G.H. Hardy and the German physician, W. Weinberg and is known as the Hardy-Weinberg law.

A generalized Hardy-Weinberg law: Suppose the gene pool of a panmictic population at equilibrium consists of p A + q a alleles such that p + q = 1. Following random union of gametes, the subsequent population will have the following genetic composition:

		sperm	
		p A	q a
ova	p A	p^2 AA	pq Aa
	q a	pq Aa	q^2 aa

genotypic composition = p^2AA + 2 pq Aa + q^2aa = 1

allele frequencies = p A + q a = 1

NOTES

Assume the population consisting of p^2 AA + 2pq Aa + q^2 aa is allowed to mate at random. What is the predicted genetic composition of the next generation? This is computed as follows:

Mating	Frequency of Mating	Frequency of Progeny: AA	Aa	aa
AA x AA	$p^2 \times p^2 = p^4$	p^4		
AA x Aa	$2(p^2 \times 2pq) = 4p^3q$	$2p^3q$	$2p^3q$	
AA x aa	$2(p^2 \times q^2) = 2p^2q^2$		$2p^2q^2$	
Aa x Aa	$2pq \times 2pq = 4p^2q^2$	p^2q^2	$2p^2q^2$	p^2q^2
Aa x aa	$2(2pq \times q^2) = 4pq^3$		$2pq^3$	$2pq^3$
aa x aa	$q^2 \times q^2 = q^4$			q^4
	TOTAL	$p^4+2p^3q+p^2q^2$	$2p^3q+4p^2q^2+2pq^3$	$p^2q^2+2pq^3+q^4$
	simplies to	p^2	$2pq$	q^2

Thus, at equilibrium, a population with a gene pool of pA + qa=1 will consist of p^2 AA + 2pq Aa + q^2 aa=1 individuals. This is the generalized Hardy-Weinberg equation.

Verification of random mating in human populations: Any meaningful application of the Hardy-Weinberg law to human populations depends upon demonstrating that such populations are panmictic. One such test of panmixia was made using the M-MN-N blood types in a Japanese population. The population consisted of 1,482 individuals distributed as follows:

M-N type	Genotype	Observed Number	Computed Allele Frequency	Expected Number
M	a^Ma^M	406	$a^M = \frac{406 + 372}{1482} = 0.525$	408.44
MN	a^Ma^N	744	$a^N = \frac{332 + 372}{1482} = 0.475$	739.22
N	a^Na^N	332		334.34
		1,482		1,482.00

NOTES

In actual fact, the 1,482 individuals consist of 741 couples whose M-N blood type was determined as part of their marriage registration. The observed frequencies of all six possible marriages are given in the table below. In the same table the expected frequencies of these marriages are listed assuming panmixia prevails. These are computed by letting p equal the frequency of allele a^M and q equal the frequency of allele a^N. By substituting the finite values of 0.525 for p and 0.475 for q the expected numbers of each marriage has been calculated. Thus it is possible to compare observation with expectation.

Marriage type	Observed number	Expected proportion	Expected number
M X M	58	$p^2 \text{ X } p^2 = p^4 = 0.0760$	56.3
M X MN	202	$2(p^2 \text{ X } 2pq) = 4p^3q = 0.2749$	203.7
MN X MN	190	$2pq \text{ X } 2pq = 4p^2q^2 = 0.2487$	184.3
M X N	88	$2(p^2 \text{ X } q^2) = 2p^2q^2 = 0.1244$	92.2
MN X N	162	$2(2pq \text{ X } q^2) = 4pq^3 = 0.2251$	166.8
N X N	41	$q^2 \text{ X } q^2 = q^4 = 0.0509$	37.7
TOTAL	741	1.0000	741.0

Note that the observed and expected numbers of the several kinds of marriages are in good agreement. It is concluded that so far as the M-N blood types are concerned, this population is panmictic.

The application of the Hardy-Weinberg law to genetic analysis: The Hardy-Weinberg law can be used to test the hypothesis that a universal trait is conditioned by a recessive, autosomal gene. The method is especially useful because only three pieces of information are needed: (1) the phenotypes of the parents, (2) the phenotypes of all their children, and (3) an estimate of the allelic gene frequencies in the population.

To illustrate the method, the genetics of tasting and non-tasting the chemical compound, phenylthiocarbamide (PTC), a rat poison, will be analyzed. People in any human population as well as some higher primates can be classified into two classes: tasters of PTC and non-tasters of PTC. In the U.S. it was found that the Caucasian population consists of 70.2% tasters and 29.8% non-tasters. A study of kindreds suggested that tasting is probably conditioned by a dominant allele T and non-tasting by the recessive allele t. Since genetic conclusions from kindred studies must be tempered with caution, can an independent determination of the genetic basis of PTC tasting and

NOTES

non-tasting be made? It is wholly realistic to assume that neither the ability to taste nor not taste PTC is a precondition for marriage. Hence, it can be assumed that so far as tasting is concerned, marriages occur entirely at random in the population. A sample population analysis of the U.S. Caucasian population gave these results:

Phenotype parents	Number families	Children Taster	Children Non-taster	Proportion children who are non-tasters
Taster X Taster	452	929	130	0.123
Taster X Non-taster	289	483	278	0.336
Non-taster X Non-taster	86	5	218	0.978

If three assumptions are made: (1) the autosomal alleles T and t condition the two phenotypes, (2) panmixia prevails, and (3) the population is at genetic equilibrium, then the genetic composition of the population is described by expression $p^2 \underline{TT} + 2pq\ \underline{Tt} + q^2 \underline{tt} = 1$. If this relationship holds, it should be possible to compute for each of the three kinds of marriages the expected proportion of their progeny who do not taste PTC. A comparison of the observed and expected proportions of non-tasters serves to test the hypothesis that the phenotypes are conditioned by segregation of the autosomal alleles T and t. In table 3 the expressions for the expected proportion of non-taster progeny for each marriage have been derived in terms of q, the frequency of the allele t for any panmictic population at equilibrium. Calculation of the expected frequency of non-tasters from marriages of non-tasters X non-tasters is straight forward. This is the mating tt X tt and all the progeny are expected to be non-tasters. To compute the proportion of non-taster progeny from marriages of taster X taster and taster X non-taster an estimate is needed for the value q, the frequency of the t gene in the Caucasian population. As noted earlier, non-tasters constitute 29.8% of the U.S. Caucasian population. If non-tasters are homozygous tt, and if the population is panmictic at genetic equilibrium, then 29.8% equals $q^2 \underline{tt}$, and $q_{\underline{t}}$ is the square root of 29.8% or 0.545. By substituting the t allele frequency into the appropriate expressions, the desired comparisons between the observed and expected proportions of non-taster progeny for all three matings can be made:

Table 3. Expected non-taster progeny after random marriages among PTC taster (T-) and non-taster (tt) individuals. (frequency gene T = p, gene t = q. p + q = 1.)

Marriage	Genotype Parents	Frequency Marriage	Frequency of genotype TT	Tt	tt
1) Taster X Taster	TT X TT	$(p^2)(p^2) = p^4$	p^4		
	TT X Tt	$2(p^2)(2pq) = 4p^3q$	$2p^3q$	$2p^3q$	
	Tt X Tt	$(2pq)(2pq) = 4p^2q^2$	p^2q^2	$2p^2q^2$	p^2q^2
2) Taster X Non-taster	TT X tt	$2(p^2)(q^2) = 2p^2q^2$		$2p^2q^2$	
	Tt X tt	$2(2pq)(q^2) = 4pq^3$		$2pq^3$	$2pq^3$
3) Non-taster X Non-taster	tt X tt	$(q^2)(q^2) = q^4$			q^4

Proportion (f_1) of tt progeny from Taster X Taster marriages =

$$f_1 = \frac{p^2q^2}{p^4 + 4p^3q + 4p^2q^2} = \frac{p^2}{p^2}\left(\frac{q^2}{p^2 + 4pq + 4q^2}\right) = \left(\frac{q}{p + 2q}\right)^2 = \left(\frac{q}{1 + q}\right)^2$$

Proportion (f_2) of tt progeny from Taster X Non-taster marriages = $f_2 = \frac{2pq^3}{2p^2q^2 + 4pq^3} = \frac{2pq^2}{2pq^2}\left(\frac{q}{p + 2q}\right) = \frac{q}{1 + q}$

Proportion (f_3) of tt progeny from Non-taster X Non-taster marriages = $f_3 = \frac{q^4}{q^4} = 1$

	Proportion of non-taster progeny	
Parents	Observed	Expected
Taster X Taster	0.123	$\left(\frac{0.545}{1.545}\right)^2 = 0.124$
Taster X Non-taster	0.336	$\frac{0.545}{1.545} = 0.352$
Non-taster X Non-taster	0.978	1.0

Agreement between observed and expected is excellent. It is, therefore, reasonable to conclude that the occurrence of the phenotypes tasting or non-tasting of PTC is best explained by segregation of the alleles $\underline{T}$ and $\underline{t}$ in a panmictic population at genetic equilibrium.

It can be argued that these results are fortuitous and do not represent the genetic basis of PTC tasting. A test of the correctness of the conclusions can be made by examining a completely independent population with quite different frequencies of tasters and non-tasters. This has been done for American Indians where the tasters make up over 90% of the population. The results confirm fully the conclusions based upon the U.S. Caucasian population.

The ratio of homozygotes to heterozygotes in a population: In the case of rare traits where panmixia can be assumed, application of the Hardy-Weinberg law provides insights into the genetic structure of the population. Consider the case of albinism, conditioned by an autosomal, recessive gene and occurring in a wide variety of ethnic groups. In several European populations, albinos occur in about 1/20,000 individuals. If panmixia is assumed, the frequency of the albino gene is computed as the square root of 1/20,000 or ca. 1/140. Thus, the frequency of heterozygotes approximates 2pq or 2(139/140)(1/140) or ca. 1/70. The significance of this simple calculation is self-evident. Whereas the albinos are comparatively rare, heterozygotes are quite common, and the bulk of the albino genes are carried and transmitted by heterozygous individuals not homozygous ones. This relationship is further emphasized in the accompanying table where a number of gene frequencies are given. If albinism is considered to be an undesirable trait and negative measures are practiced to eliminate the albino gene, it

Table 4a. Change in frequency of the recessive gene a in response to complete selection against homozygous aa individuals.

Generation	Frequency of genotypes			Frequency of genes	
	AA	Aa	aa	a	A
0 Before selection	p^2	2pq	q^2	q	p
After selection	$\frac{p^2}{p^2 + 2pq}$	$\frac{2pq}{p^2 + 2pq}$	0	$\frac{pq}{p^2 + 2pq} = \frac{q}{1 + q}$	$\frac{p^2 + pq}{p^2 + 2pq} = \frac{1}{1 + q}$
1 Before selection	$\frac{1}{(1+q)^2}$	$\frac{2q}{(1+q)^2}$	$\frac{q^2}{(1+q)^2}$	$\frac{q}{1 + q}$	$\frac{1}{1 + q}$
After selection	$\frac{\frac{1}{(1+q)^2}}{\frac{1 + 2q}{(1+q)^2}}$	$\frac{\frac{2q}{(1+q)^2}}{\frac{1 + 2q}{(1+q)^2}}$	0	$\frac{\frac{q}{(1+q)^2}}{\frac{1 + 2q}{(1+q)^2}} = \frac{q}{1 + 2q}$	$\frac{\frac{1 + q}{(1+q)^2}}{\frac{1 + 2q}{(1+q)^2}} = \frac{1 + q}{1 + 2q}$
2 Before selection	$\frac{(1+q)^2}{(1+2q)^2}$	$\frac{2q(1+q)}{(1+2q)^2}$	$\frac{q^2}{(1+2q)^2}$	$\frac{q}{(1+2q)}$	$\frac{1 + q}{1 + 2q}$
After selection	$\frac{\frac{(1+q)^2}{(1+2q)^2}}{\frac{(1+q)^2 + 2q(1+q)}{(1+2q)^2}}$	$\frac{\frac{2q(1+q)}{(1+2q)^2}}{\frac{(1+q)^2 + 2q(1+q)}{(1+2q)^2}}$	0	$\frac{\frac{q(1+q)}{(1+2q)^2}}{\frac{(1+q)^2 + 2q(1+q)}{(1+2q)^2}} = \frac{q}{1 + 3q}$	$\frac{\frac{(1+q)^2 + q(1+q)}{(1+2q)^2}}{\frac{(1+q)^2 + 2q(1+q)}{(1+2q)^2}} = \frac{1+2q}{1+3q}$
n Before selection				$\frac{q}{1 + nq}$	$\frac{1 + (n - 1)q}{1 + nq}$

Relative frequencies of individuals homozygous and heterozygous for a recessive gene at different gene frequencies

gene frequency %	50	10	3.16	1.41	1	0.707	0.316	0.10
% homozygotes	25	1	0.1	0.02	0.01	0.005	0.001	0.0001
% heterozygotes	50	18	6.12	2.79	1.98	1.40	0.63	0.20
no. heterozygotes to one homozygote	2	18	61	140	198	281	630	2,000

is obvious that elimination from the breeding population of the albinos by whatever means is neither an efficient nor an effective means of reducing the albino gene frequency. Clearly, effective measures depend upon identifying the heterozygotes and minimizing their contribution to the gene pool of the next generation.

8. Selection and Population Structure

To illustrate the effects of selection on gene frequency, consider a panmictic population consisting of p^2 AA + 2pq Aa + q^2 aa individuals. Suppose aa individuals are kept from breeding and thereby contribute none of their genes to the next generation. The AA and Aa individuals breed at random. What is the effect on the a allele and on the homozygous aa frequencies in the subsequent generation? As shown in table 4a, by such drastic selection the frequency of gene a changes from q to (q/1+q) and panmixia of the residual population results in homozygous aa individuals in a frequency of $(q/1+q)^2$. Continued practice of complete selection against homozygous aa individuals in generation 1 further reduces the a gene frequency from (q/1+q) to (q/1+2q) and leads to a frequency of $(q/1+2q)^2$ homozygous aa individuals in the next generation. At the end of n generations of such selection, the recessive allele frequency is (q/1+nq). The efficacy of this method of selection is outlined in table 4b where three quite different recessive gene frequencies are compared. Note that where the gene frequency is relatively high, selection against the homozygous recessive is a comparatively effective way of eliminating the recessive gene. On the contrary with a low gene frequency, selection against the homozygous recessive has an essentially trivial effect in decreasing the recessive gene frequency. Thus, for the homozygous recessive trait of low incidence selection against the homozygotes is a relatively inefficient and uneconomic way of eradicating the trait. The efficient elimination of such traits is

Table 4b. Elimination of homozygous recessive individuals and the recessive gene frequency.

Generation	Recessive gene frequency		
0	0.5	0.1	0.01
1	0.33	0.091	0.0099
2	0.25	0.083	0.0098
3	0.2	0.077	0.0097
4	0.167	0.071	0.0096
5	0.143	0.067	0.0095
–			
10	0.083	0.05	0.0090

Question: How long will it take (generations) to reduce the PKU frequency from 1/25,000 to 1/50,000?

Let q_o = the original gene frequency = $\sqrt{1/25,000}$

q_n = the ultimate gene frequency = $\sqrt{1/50,000}$ after n generations

$$q_n = \frac{q_o}{1+nq_o} \text{ or } q_o = q_n + nq_oq_n$$

$$n = \frac{q_o}{q_oq_n} - \frac{q_n}{q_oq_n} = \frac{1}{q_n} - \frac{1}{q_o} = \sqrt{1/_{1/50,00}} - \sqrt{1/_{1/25,000}}$$

$$n = \frac{1}{\frac{1}{225}} - \frac{1}{\frac{1}{158}} = 225 - 158 = 67 \text{ generations}$$

if a human generation = 25 years, time = 67 X 25 = 1675 years.

predicated upon identifying the heterozygotes and minimizing or restricting their contribution to the gene pool of the succeeding generation.

9. Migration and the Genetic Structure of Populations

A condition necessary for the maintenance of Hardy-Weinberg equilibrium is that the population be for all practical purposes reproductively closed. This means that no new genes are brought into the population by immigrants from an unrelated population and no genes are lost by emigration. Should immigrants enter the population, their contribution of genes to the population gene pool could upset the equilibrium and bring about a significant change in the frequency of genes in the population. The role of migration on the genetic structure of a human population can be illustrated by considering miscegenation as the basis for the migration of alleles from one population to another population. Subsequent to the introduction of slavery into the U.S., primarily during the 18th century, miscegenation led to a migration of genes from the white population to the black population. The extent to which this migration altered the genetic composition of the black population can be estimated by computing M, the present proportion of genes at a locus in the black population derived from the white population. The accuracy of this estimate depends upon the fulfillment of these criteria: (1) that precise gene compositions of the ancestral populations, African black and European white, are known; (2) that no change has occurred in the gene frequencies of the ancestral and contemporary populations; (3) that interbreeding of the ancestral populations, slave and white, is the only factor affecting the gene frequencies of the U.S. blacks; and (4) that adequate, unbiased samples of all three contemporary populations, U.S. white, U.S. black, and West African black, are available. Assuming all criteria are fulfilled--at best an obvious oversimplification--M can be estimated most accurately where a particular gene occurs in a low frequency in U.S. white population (q_c) and in a high frequency in the African black population (q_a) or vice versa. After determining the frequency of the gene in the U.S. black population (q_n), M can be estimated from the equation derived in table 4c. This can be illustrated as follows. Among the several alleles at the Rh blood group locus, one allele, Rh_0, is comparatively infrequent among U.S. whites ($q_c = 2.2\%$) and comparatively frequent among West African blacks ($q_a = 63.4\%$). The frequency of the Rh_0 gene among the U.S. blacks, q_n, was found to be 43.8%. Using the three Rh_0 gene frequencies and making the appropriate numerical substitutions,

Table 4c. Miscegenation as the basis of gene migration in the U.S. black population.

Let M = present frequency of genes at a locus derived from the white population

q_a = gene frequency among African ancestors

q_n = gene frequency among U.S. blacks

q_c = gene frequency among white ancestors

$$q_n = Mq_c + (1-M)q_a$$

$$q_n = Mq_c + q_a - Mq_a$$

$$q_n - q_a = Mq_c - Mq_a$$

$$M = \frac{q_n - q_a}{q_c - q_a}$$

$$M = \frac{0.438 - 0.634}{0.022 - 0.634} = 0.31 \text{ or } 31\%.$$

This means that if the frequency of the Rh_0 is representative of all genes in the U.S. black population, 31% of the genes are of white origin, presumably derived by miscegenation.

10. Consanguinity

In consanguineous marriages or mating between relatives, inbreeding occurs with the concomittant consequence of increased homozygosity for recessive genes. Very likely this is the empirical reason underlying both religious and secular prohibitions against marriages between relatives since often the homozygous recessive genotype determines an undesirable trait. That in practice this is the case is emphasized by the accompanying table in which the relationship between first cousin marriages and the occurrence of several hereditary diseases is given for Caucasian and Japanese populations.

To emphasize the increased likelihood of homozygosity for recessive genes incidental to consanguinity, consider figure 7, an hypothetical kindred. Assume person III-1 has the genotype Aa where a is a rare recessive autosomal gene. What is the probability that III-2, the first cousin, is also heterozygous Aa? The question can be rephrased as follows: What is the probability that III-2 inherited the same a allele as III-1?

Person III-1 inherited a either from parent II-1 or parent II-2 with equal probability or $\frac{1}{2}$. If II-2 carries a, he inherited a from either I-3 or I-4. I-3 or I-4 each has a chance of $\frac{1}{2}$ of transmitting a. Thus, the chance that II-3 carries a equals $\frac{1}{2} \times \frac{1}{2}$ or $\frac{1}{4}$. The chance that II-3 will transmit allele a to III-2 is $\frac{1}{2}$; hence the overall probability that III-2 possesses allele a is $\frac{1}{2} \times \frac{1}{4}$ or $\frac{1}{8}$. Therefore, if an individual is Aa, the chance his first cousin also is Aa is $\frac{1}{8}$. The significance of this relationship can be best understood if a comparison is made between the $\frac{1}{8}$ chance of heterozygosity of the first cousin and the chance of heterozygosity of an unrelated individual drawn at random from a population. Assume the population is panmictic at genetic equilibrium. The heterozygosity of an unrelated individual is a function of q, the population frequency of allele a. If allele a is as rare as $\frac{1}{50}$, the chance the unrelated individual is heterozygous becomes 2pq or ca. $\frac{1}{25}$, or if a is an infrequent as $\frac{1}{200}$, 2pq equals ca. $\frac{1}{100}$. Thus, the rarer the allele a, the less likely an Aa individual will marry another Aa individual drawn at random from the population.

Figure 7. Hypothetical kindred with a first cousin marriage

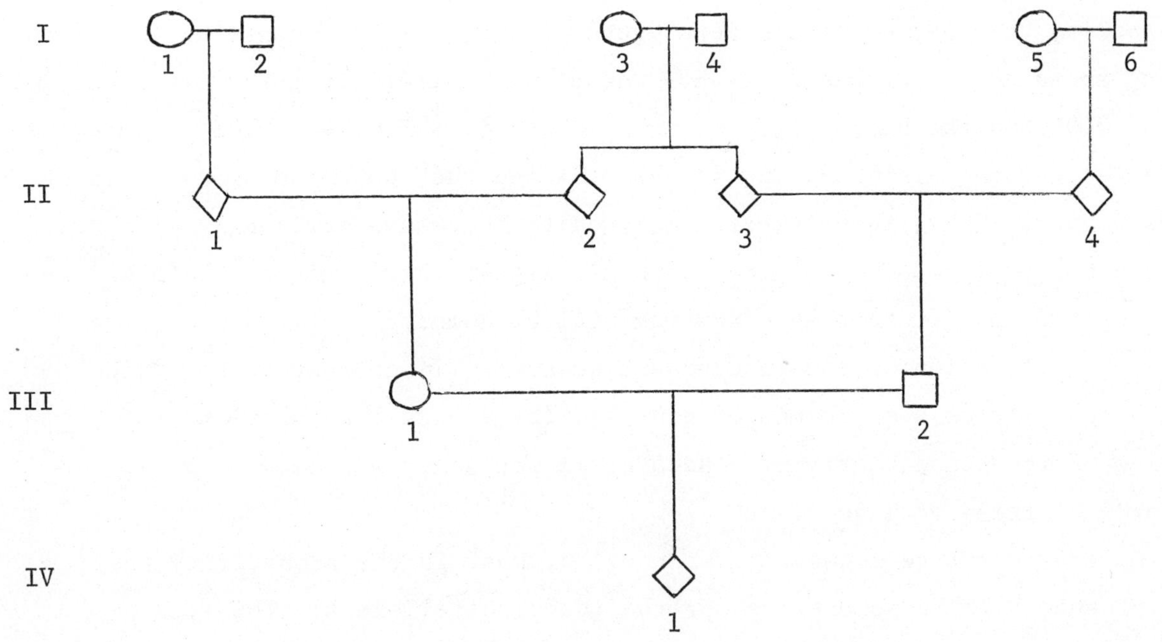

Consanguinity among parents of individuals with a recessive inherited disease. C = Caucasian, J = Japanese populations (after Neel and Schull 1954).

Trait		Incidence of first cousin marriages among parents: This Disease	General Population
Albinism	C	0.18-0.24	0.01
	J	0.37-0.59	0.06
Infantile amaurotic idiocy	C	0.27-0.53	
	J	0.55-0.85	
Ichthyosis congenita	C	0.30-0.40	
	J	0.67-0.93	
Congenital total colorblindness	C	0.11-0.21	
	J	0.39-0.51	
Xeroderma pigmentosa	C	0.20-0.26	
	J	0.37-0.43	

From these considerations, it follows that the expectation of an aa child changes drastically with the allele frequency. The first cousin marriage between III-1, known to be Aa, and III-2 means that the probability IV-1 will be aa is $\frac{1}{8}$ X $\frac{1}{4}$ or $\frac{1}{32}$. However, if III-1 marries an unrelated person, the probability this person is a heterozygote is 2pq, and the chance of their producing an aa child becomes 2pq X $\frac{1}{4}$. Where the a allele frequency is 0.02, 2pq is ca. 0.04 and the probability of an aa child is $\frac{1}{4}$ X 0.04 = 0.01. Where the a allele frequency is 0.02, 2pq is ca. 0.01 and the chance of an aa child becomes 0.0025. Thus, the risk of homozygosity decreases strikingly with outcrossing.

The frequency of homozygotes in cousin marriages: For a given cousin marriage, the probability that an offspring will be homozygous for a recessive gene a is a function of two independent events: (1) consanguinity, which unites two alleles a of common origin, and (2) chance homozygosity of two a alleles of independent origin. The overall frequency will equal the sum of the probabilities of each event.

Given a marriage between first cousins, what is the probability they will produce a child homozygous for the identical allele a? This can be followed by referring to the hypothetical kindred. If the allele a is carried by one of the ancestors of IV-1, e. g., I-3, the probability of a being transmitted to IV-1 via II-2 and III-1 is $\frac{1}{2}$ X $\frac{1}{2}$ X $\frac{1}{2}$ or $\frac{1}{8}$. The probability of a being transmitted to IV-1 via the alternative path of II-3 and III-2 is also $\frac{1}{2}$ X $\frac{1}{2}$ X $\frac{1}{2}$ or $\frac{1}{8}$. Combining the two pathways, the probability IV-1 is homozygous aa is $\frac{1}{8}$ X $\frac{1}{8}$ or $\frac{1}{64}$. In actuality, homozygosis for a is a function of its population frequency or q. Thus, homozygosis for a specific allele of common origin is the product $(\frac{1}{64})q$. Since the great grandparents of IV-1, I-3 and I-4, between them have a total of four alleles at the a locus with each allele having a probability of $(\frac{1}{64})q$ of becoming homozygous, the overall probability of homozygosis for aa becomes 4 X $(\frac{1}{64})q$ or $(\frac{1}{16})q$.

The frequency of aa individuals, where the a alleles are of independent origin, can be seen from the following consideration. If $(\frac{1}{16})q$ is the probability that IV-1 is homozygous for the same a allele, then $(\frac{15}{16})q$ is the probability of not being homozygous for the same allele. Since the population frequency of an a allele of independent origin is q, the chance of combining the two a alleles of independent origin is $(\frac{15}{16})q$ X q or $(\frac{15}{16})q^2$. Thus, the overall probability that a first cousin marriage will result in a homozygous aa child is $(\frac{1}{16})q + (\frac{15}{16})q^2$ or $(\frac{q}{16})(1 + 15q)$.

It is now possible to compare homozygosis frequencies for a recessive allele in a panmictic population and in first cousin marriages. Three allele frequencies are given in table 5 together with the computed frequencies of

NOTES

homozygous recessive progeny expected for random and first cousin marriages for each allele frequency. Note that where the allele frequency is high, there is little difference in the frequency of homozygous progeny derived from first cousin and random marriages. However, where the allele frequency is low (0.01) but not especially rare, homozygous progeny are 7 times more frequent in cousin marriages than in random marriages. There is little wonder that undesirable traits, the result of homozygous recessive genes, are relatively common among the progeny of cousin marraiges.

11. Genetic Drift and the Genetic Structure of Populations

If a population is to remain in Hardy-Weinberg equilibrium from one generation to the next, not only must the population, and therefore its gene pool, be extremely large, but the sample of zygotes drawn from the gene pool to make up the next generation must be a random sample and thus be representative of the entire population. If the sample of zygotes drawn from the parental population is non-representative, deviations from equilibrium can occur either as chance fluctuations in the gene frequencies or as chance fixation of a gene. This is genetic drift. Genetic drift depends *per se* on the effective breeding size of a population. It can occur either when a population undergoes periodic drastic fluctuations in size, the so-called bottleneck effect or when a colonizing population at its inception is but a small sample of its ancestral population, the so-called founder effect.

A probable case of genetic drift is the Dunker religious isolate in the U.S. This group originated from the German Baptist Brethern sect founded in 1708 at Krefeld, Germany. In 1719, 28 members of the sect migrated to eastern Pennsylvania and from these immigrants the present members of the sect descended. In table 5a are listed the blood group gene frequencies of the German population in the neighborhood of Krefeld, the U.S. population, and the isolated Dunker population. Note that while the U.S. and German populations do not appreciably differ in their blood group gene frequencies, the blood group gene frequencies among the Dunkers differ strikingly from both. This difference is interpreted as an instance of genetic drift. The small number of immigrants, founders of the contemporary U.S. community, were a non-representative sample of the blood group genes in the original population and thus the differences in gene frequencies.

Genetic drift has similarly been invoked to explain the significant differences in blood group frequencies occurring among Amerindians and among Eskimos.

Table 5. Frequency of homozygous recessive children from first cousin and random marriages

recessive allele frequency, q	homozygous progeny from random marriages = q^2	homozygous progeny from cousin marriages = $\frac{q}{16}(1+15q)$
0.5	25% or 0.25	26.5% or 0.265
0.1	1% or 0.01	1.56% or 0.0156
0.01	0.01% or 0.0001	0.072% or 0.00072

Table 5a.

Blood group gene frequencies in the Dunker community

	Blood Group Gene Frequencies				
Population	I^a	I^b	i	a^M	a^N
Germany	0.286	0.074	0.639	0.548	0.452
Dunker	0.378	0.025	0.597	0.655	0.345
U.S.	0.258	0.041	0.701	0.540	0.460

Founder effects: The chance effects of a small number of parents on the gene pool of their descendants has been termed the founder principle. In effect, an undesirable trait is introduced into a small interbreeding population by one or a few individuals and remains in the population for an indefinite period. A number of examples of this type have been recognized in human populations of which one hereditary disease, variegate porphuria, is an especially striking case. This is a metabolic disorder associated with impaired porphyrin metabolism and conditioned by a Mendelian dominant gene. Symptoms include skin photosensitivity and acute events with pain, vomiting, constipation, and paralysis plus psychiatric manifestations. Variegate means some individuals have skin symptoms, others with acute symptoms and still others with both. There are over 10,000 cases of variegate porphyria in South Africa, practically all traced to two Dutch settlers who married there in 1688.

Limitations of Hardy-Weinberg: Random mating is in all probability rarely fulfilled in human populations primarily because varying degrees of consanguinity occur. In some cultures consanguineous marriages are frequent. In other cultures assortative matings occur. Both systems of matings are departures from panmixia, and their effects on population structure should not be minimized.

A closer examination of human populations reveals that most are not uniform but rather are composed of subgroups or isolates. Two kinds of isolates occur: (1) geographical, (2) social. In general, population isolates were comparatively separate groups which are only slowly intermixing. The prejudices against intermixing are well-known. Subsequently, further consideration will be given to this subject.

12. Lethal and Sublethal Gene Effects

The spectrum of phenotypic effects produced by various gene substitutions is broad. On the one hand homozygosity for a particular gene, e. g., the recessive alkaptonuria gene, results in a striking phenotype but with little or no effect on the viability of affected individuals. On the other

Table 6.

Results of crosses between short-tailed and normal mice

Cross	Phenotypes and Frequency of Progeny			
	Brachy	Normal	Σ	Mean Litter size
Brachy X Normal	934 = 46% (1016 = 50%)*	1098 = 54% (1016 = 50%)*	2032	5
Brachy X Brachy	3639 = 65% (4189.5 = 75%)*	1947 = 35% (1396.5 = 25%)*	5585	4

*in parentheses the Mendelian expectations for each class are given.

hand, homozygosity for quite another recessive gene leads to an alteration in a vital process so severe that the affected individual dies. For purposes of discussion, three more or less arbitrary classes of genes can be recognized which are directly responsible for the death of an affected individual. (1) Lethal genes: those which impair prenatal development and the affected individual does not survive to birth. (2) Sublethal genes: those which cause the demise of the affected individual after birth but before reproductive age. (3) Delayed lethal genes: those which cause the death of affected individuals after reproductive age.

Crosses with short-tailed (brachyury) mice illustrate a model system of lethal gene inheritance. The cumulative results of a number of crosses of Brachy X Brachy and Brachy X Normal are presented in table 6. From the cross Brachy X Normal, two conclusions are possible: Brachyury is conditioned by a dominant allele and the results are compatible with a Mendelian test cross segregation. Thus, brachyury individuals are heterozygous (B/b), and normal mice are homozygous (b/b). These conclusions are, however, not entirely compatible with the results of the Brachy X Brachy cross. The frequencies of the resultant progeny deviate significantly from Mendelian expectation; there is a significant excess of normal and deficiency of brachy mice. A clue to the reason for this discrepancy can be found in the mean litter size in the two crosses. The Brachy X Brachy cross produced 20% fewer progeny than did the Brachy X Normal cross. A careful examination of the developing embryos derived from the Brachy X Brachy cross provided an explanation for the discrepancy. At 8.5-9 days in development, ca. 25% of the embryos appear abnormal and their neural tubes do not differentiate normally. By 10-11 days development these abnormal embryos die and are resorbed. This and other information make is reasonable to assume that the lethal embryos are B/B in genotype. Therefore, in the cross Brachy X Brachy Mendelian segregation does occur but is not realized because of the lethal effect of the homozygous Brachy gene. The loss of the B/B progeny means that the expected progeny should consist of 67% Brachy and 33% Normal mice. As indicated in table 6, results consistant with this prediction were found.

A trait in man which appears to parallel the transmission and lethal effects of brachyury mice is brachydactyly, a shortening of the middle phalanx of the index fingers and the second toes. The kindred in figure 8a is essentially self-explanatory. The trait occurs in both sexes, is transmitted from one generation to the next and appears to be lethal when homozygous as judged by one lethal individual stemming from a marriage between two brachydactylous individuals.

Figure 8a.

Brachydactyly = shortening of middle phalanx index finger and second toe where homozygote possibly lethal.

(after Mohr and Wriedt, 1919)

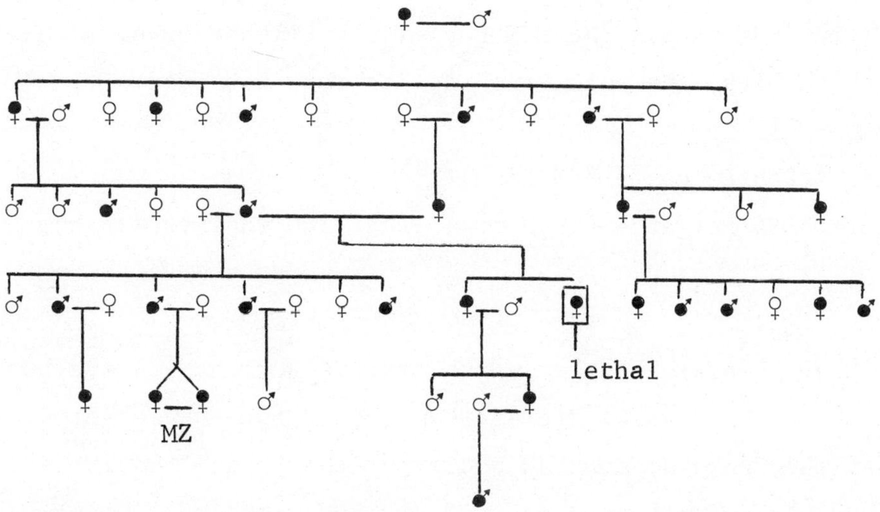

●♂ = brachydactyly

♀♂ = normal

Figure 8b. A Kindred of the Pelger Anomoly in Man

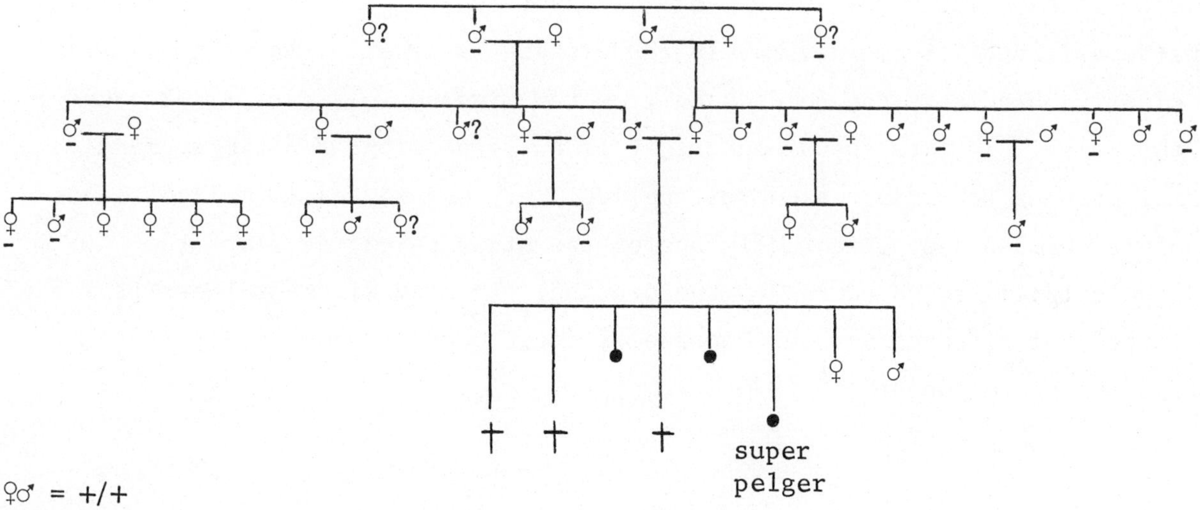

♀♂ = +/+

♀♂ (underlined) = Pg/+

† = lethal infancy

● = abortions

The socalled Pelger anomaly of rabbits and man illustrates both the inheritance of a sublethal gene and the usefulness of studies with experimental animals to human heredity. The Pelger phenotype of rabbits and man involves the polymorphonuclear leucocytes (PMN): the neutrophils, basophils and eosinophils. In peripheral blood the nuclei of PMNs are normally multilobed. In Pelger individuals the PMN nuclei are bilobed or horseshoe-shaped. In crosses with rabbits, the following results were obtained:

Pelger X Normal $\rightarrow$ 268 Pelger + 285 Normal
Pelger X Pelger $\rightarrow$ 484 Pelger + 249 Normal + ca. 200 individuals who die shortly after birth.

These data are explained by assuming that Pelger rabbits are heterozygotes and those rabbits which die are homozygous for the sublethal Pelger gene. Interestingly enough, the PMN nuclei of the lethal rabbits are completely round without any evidence of lobular structure. Occasionally a homozygote survives but is stunted with bilateral microphtholmia.

Figure 8b, a kindred of the Pelger anomaly in man, shows a parallel genetic situation and suggests that the genetic interpretation for the rabbit applies equally well to man.

Huntington's chorea is a disease characterized by the loss of voluntary muscular control and progressive mental deterioration and is an example of a delayed lethal gene. The average age of onset is about 36 years but there is considerable variation. Ultimately the disease is fatal. That the disease is clearly hereditary can be deduced from the kindred in figure 9. Because of the relatively late age of onset, it is quite possible for two seemingly normal persons to produce a choreic child, i.e., one who is normal but later develops the disease. Eventually one or the other parent develops the disease symptoms and succumbs to the disease. In many kindreds there is a deficiency of affected persons because of their death through other causes before the age of onset.

13. Sex Linkage

The facts of Mendelian genetics considered thusfar concern transmission of genes localized to the autosomes. That the autosomes are involved requires the use of hindsight for Mendelian genetics as such gives little

Figure 9. A Kindred of Huntington's Chorea

(after Spillane and Phillips, 1937)

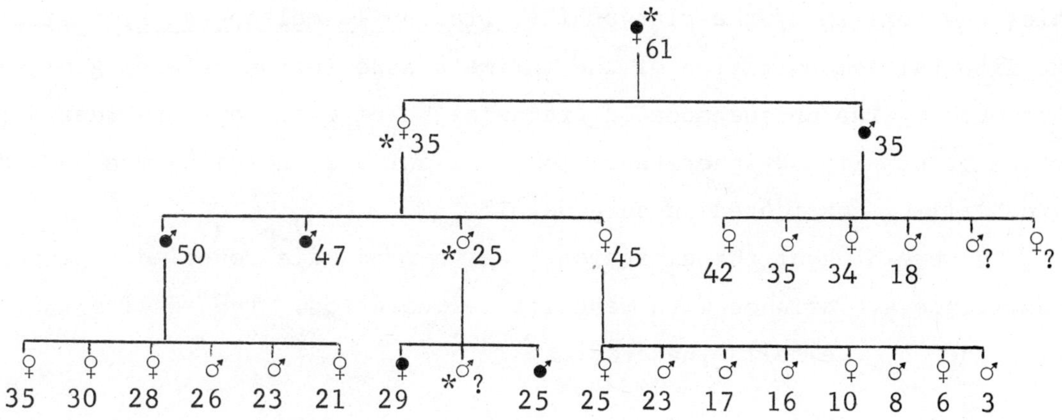

♂ ♀ (filled) = Huntington's chorea

*♂ *♀ = individuals dead at time of investigation

The number accompanying each ♂ and ♀ is their age at time of investigation or at death.

direct information associating genes with chromosomes and no direct informating linking particular genes to specific chromosomes. The discovery in 1910 by T.H. Morgan of a single white-eyed male in a culture of red-eyed males and females of the vinegar fly, *Drosophila melanogaster*, initiated the experimental demonstration of the intimate association between gene and chromosome. The unique mode of transmission of white eye provides a model system of sex-linked inheritance for most animals including man and, therefore, will be considered in some detail.

The breeding of the exceptional white-eyed male revealed a pattern of inheritance at variance with Mendelian expectation. The results are summarized by the following crosses:

white-eyed ♂ X red-eyed ♀ $\rightarrow F_1$ ♂♀ all red-eyed

F_1 X $F_1 \rightarrow F_2$ ♀♀ all red-eyed; ♂♂ ca. 1/2 red-eyed, 1/2 white-eyed.

It should be noted that in *Drosophila* (as in man) the male is the heterogametic sex possessing an XY sex chromosome complement in contrast to the homogametic or XX female. This fact plus three simple assumptions led to a reasonable and testable explanation of the unusual white-eye segregation. The assumptions are: (1) the gene for white-eye color (*w*) is recessive to red (*W*), (2) *W* and *w* are linked to the X chromosome, and (3) the Y chromosome carries no allele of white, neither *W* nor *w*. In short, the inheritance of white eye is assumed to be inseparably associated with the transmission of the X chromosomes. The regular disjunction of the sex chromosomes in males and females leads to precisely the observed genetic segregation, and the X chromosome of the exceptional white-eye male should be recovered in half his grandsons; the other half should receive a red-eye determining X. Furthermore, the recessivity of white eye means all the granddaughters should be red-eyed with half heterozygous for the white-eye allele. Progeny tests established the correctness of this prediction. Linkage of white eye to the X chromosome, i.e., sex-linked inheritance, leads to yet another prediction: the reciprocal cross of white-eyed females X red-eyed males should produce quite another phenotypic segregation, one which differs from the segregation derived from the cross white-eyed males X red-eyed females. The theoretical expectations which were confirmed by the actual crosses are summarized as follows:

NOTES

white-eyed ♀ X red-eyed ♂ → F_1 ♀ all red-eyed; ♂ all white-eyed

F_1 X F_1 → F_2 ♀♀ ca. 1/2 red, 1/2 white; ♂♂ ca. 1/2 red, 1/2 white.

The correlated segregation of white eye with the disjuction of the sex chromosomes is outlined in table 6a.

While the inheritance of white eye established the intimate correlation between transmission of the X chromosomes and the segregation of the W and w alleles, the unequivocal demonstration of this linkage came from yet another series of observations and experiments. As noted, the progeny of the cross white female X red male consist of red females and white males. While these expectations were generally fulfilled, infrequently two types of exceptions were found. Unexpected white-eyed females were recovered in ca. 1/2500 progeny and unexpected red-eyed males were found in ca. 1/1200 progeny. What is the origin of these exceptions? Does their occurrence invalidate the explanation of sex-linked inheritance of white eye? One possible explanation for the origin of the exceptional individuals invokes mutation. The original white-eyed male arose in a female as a rare mutation of W to w. Thus, a repeat of the same event in a red male could result in an exceptional white-eyed female. Similarily the red-eyed male exception could be explained by the occurrence of a reverse mutation, i.e., a mutation from w to W, in the parental white-eyed female. Two facts militate against the mutation hypothesis. First, the required mutation rates are enormously greater than the measured rates. Second, the observed exceptional red-eyed males and white-eyed females gave unexpected breeding results not predicted by the mutation hypothesis. The red-eyed males were invariably sterile. The white-eyed females when crossed to regular red-eyed males produced in addition to the expected red female and white male progeny numerous red-eyed males and white-eyed females. An alternative, testable explanation was proposed and experimentally examined by C.B. Bridges in 1916. Suppose that in the parental white-eyed females occasionally during anaphase of the first meiotic division the synapsed homologous X chromosomes instead of disjoining to opposite poles move together to the same pole. If the two X chromosomes are included in secondary oocyte nucleus, subsequent regular division of this cell will produce an ovum with two X chromosomes or ww in genotype. Alternatively, if the two X chromosomes are included in the first polar body

Table 6a. Sex linkage and white eye segregation

(let W = red-eyed carrying X chromosome, w = white-eyed carrying X chromosome)

1. white-eyed ♂ X red-eyed ♀ or wY X WW

   F_1 all ♂♂ ♀♀ red eyed or WY and Ww

   F_1 X F_1 or WY X Ww

   F_2 ♂♂ 1/2 white-eyed + 1/2 red-eyed or 1/2 wY + 1/2 WY

   ♀♀ all red eyed or 1/2 Ww + 1/2 WW

2. white-eyed ♀ X red-eyed ♂ or ww X WY

   F_1 all ♀♀ red-eyed or Ww; all ♂♂ white-eyed or wY

   F_1 X F_1 or wY X Ww

   F_2 ♂♂ 1/2 white-eyed + 1/2 red-eyed or 1/2 wY + 1/2 WY

   ♀♀ 1/2 white-eyed + 1/2 red-eyed or 1/2 ww + 1/2 Ww.

nucleus, the secondary oocyte will contain no X chromosome. The regular division of this secondary oocyte will produce an ovum lacking an X chromosome or nullo-X. If it is further assumed that in *Drosophila* the Y chromosome does not determine maleness *per se* but is concerned primarily with determining male fertility, the nature of the exceptional white-eyed females and red-eyed males can be predicted and tested. Suppose a two X (*ww*) ovum is fertilized by a Y-bearing sperm. The zygote, XXY or *wwY*, should develop into a white-eyed female. Alternatively, suppose the nullo-X ovum is fertilized by an X-(*W*) bearing sperm. The zygote, XO or *W*, should develop into a phenotypic male but sterile because the Y chromosome is lacking. A direct check on the correctness of this explanation entails a cytological determination of the karyotype of the exceptional males and females. In agreement with the hypothesis, Bridges found that all the white-eyed female exceptions were XXY and all the red-eyed male exceptions were XO. These findings were critical for the chromosome basis of heredity. They demonstrated that linkage of the white eye gene to the X chromosome explained both the regular transmission of white eye, and the infrequent instances of irregular transmission. Linkage of genes to specific chromosomes was proved.

The discovery of white-eyed *Drosophila* and the determination of its heredity provided an almost simultaneous explanation for the transmission of two well-known human traits, red-green colorblindness and hemophilia. Both traits had long been recognized as familial, but their precise heredity had to await the finding of white-eyed *Drosophila*.

Red-green colorblindness actually represents two discrete traits: protanopia (and the less severe protanomaly) associated with impaired perception of the red wave lengths of the visual light spectrum, and deuteranopia (and the less severe deuteranomaly) associated with impaired perception of the green wave lengths. Since their modes of transmission are identical, the two types of red-green colorblindness will be considered together. An examination of kindreds with red-green colorblindness showed that the transmission of the trait paralleled precisely the transmission of white eye in *Drosophila*. A case in point is the kindred in figure 10a. Note that the colorblindness of the grandfather reappears among his grandsons but not all his grandsons. This kindred suggests that (1) the gene for colorblindness is located on the X chromosome, (2) the Y chromosome carries no colorblindness alleles, and (3) colorblindness is recessive to normal color vision. In addition, it follows that colorblind females, all of whose sons should be colorblind, should have colorblind fathers. These deductions are

Figure 10a. A Kindred of Red-Green Colorblindness

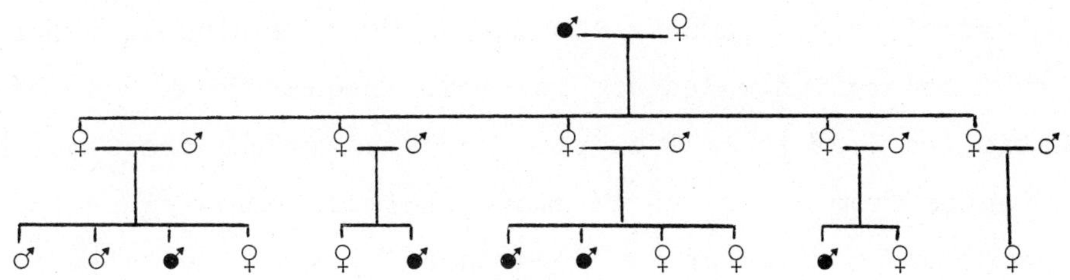

Figure 10b. A Kindred of Red-Green Colorblindness with a Colorblind Female (After Bell, 1926)

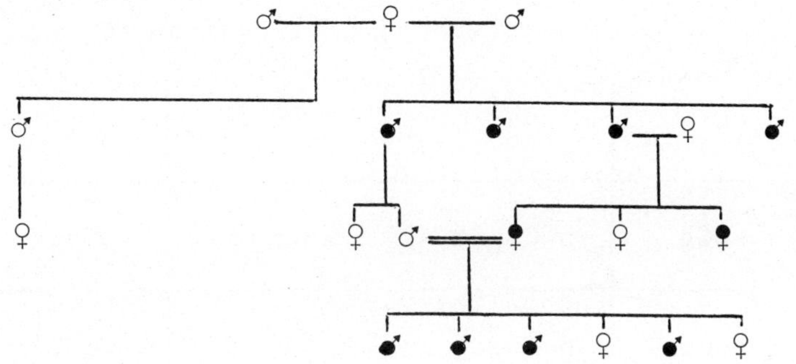

● = colorblind

♀ = normal vision

illustrated in figure 10b. Taken together, these and other data are consistent with the view that red-green colorblindness in man is recessive and sex-linked.

Sex-linkage of colorblindness provides a ready explanation for the well-known fact that colorblind males are far more frequent than colorblind females. Estimates from several ethnic groups shows the frequency of colorblind males ranges from as low as 1% among Eskimos, Melanesians and Navahos to as high as 8% among males of European ancestry. As indicated in table 7, the frequency of colorblind females is only a fraction of the colorblind male frequency. This difference can be explained in a straight-forward way by considering the population genetics of colorblindness. Since in man males are XY and females XX, the following situation obtains. Assume that a population is panmictic and that the colorblindness alleles (cb = color blind and Cb = normal vision) are at equilibrium such that the allele frequencies are $p\underline{Cb} + q\underline{cb} = 1$. What will be the genotypic and phenotypic composition of this population? Female gametes will consist of $p\underline{Cb} + q\underline{cb}$ ova; half the males gametes will consist of $p\underline{Cb} + q\underline{cb}$ X-bearing sperm and the other half will be Y-bearing sperm. The population composition is described by the following mating matrix:

		sperm		
		pCb	qcb	Y
ova	pCb	p^2Cb/Cb	pqCb/cb	pCb/Y
	qcb	pqCb/cb	q^2cb/cb	qcb/Y
		♀ progeny		♂ progeny

Thus, the females consist of the genotypes $p^2\,\underline{Cb}/\underline{Cb} + 2pq\,\underline{Cb}/\underline{cb} + q^2\,\underline{cb}/\underline{cb}$ or $p^2 + 2pq$ with normal vision and q^2 colorblind. The males consist of $p\underline{Cb}$/Y or normal vision plus $q\underline{cb}$/Y or colorblind individuals. Note that when the population is at equilibrium, the frequency of colorblind females (q^2) can be estimated by squaring the frequency of colorblind males. Thus, if an equilibrium population contains 8% colorblind males, the expected frequency of colorblind females is $8\%^2$ or 0.64%. Referring to table 7, note that the aforementioned relationship between colorblindness frequencies in males and

Table 7.

Frequency of red-green colorblindness (USA)

Ethnic group	Frequency colorblind males	Frequency colorblind females
Caucasian (1)	0.084	0.013 (0.007)*
Caucasian (2)	0.082	0.010 (0.0067)
Negro	0.039	0.008 (0.0015)
Mexican	0.023	0.006 (0.00053)

(*) = expected frequency based on male frequency.

females is approximated for all ethnic groups, but in each instance the female frequency is somewhat higher than predicted. This excess of colorblind females can be attributed to three causes: (1) occasional heterozygous females are colorblind, (2) protanopia and deuteranopia, caused by different genes, are not distinguished, and (3) some heterozygous females have "poor color aptitude."

Hemophilia, the socalled bleeder's disease, received particular notoriety through a well-known kindred, that of Queen Victoria of England and her descendants. The disease, associated with impaired blood clotting, was for many years considered unusual because it appeared only in males. A typical kindred is presented in the figure 11a. Since the blood clotting mechanism is quite complex, it is not altogether surprising that what was for a long time considered to be one disease entity is in fact two: (1) classical hemophilia or hemophilia A, constituting ca. 85% of the cases, characterized by a deficiency in the anti-hemophilia globulin, and (2) Christmas disease or hemophilia B, constituting the remaining 15% of the cases, in which the thrombloplastin factor is deficient. Both present the same pattern of inheritance and run true in kindreds. The absence of hemophiliac females was for a long time not completely understood. However, the discovery of hemophilia in dogs made the necessary controlled crosses possible, in particular the cross of a heterozygous female to a hemophiliac male. From a series of such crosses, both hemophiliac males and females and normal males and females were recovered in ca. equal frequency as expected if hemophilia is conditioned by a sex-linked recessive gene. These results suggested that human hemophiliac females had not been observed because rarely had hemophiliac males survived to a marriageable age thus obviating mating with a heterozygous female. Subsequently, the description of milder forms of hemophilia led to the discovery of human hemophiliac females as illustrated in figure 11b.

Partial sex-linkage: Genes which are located on both X and Y chromosomes are said to be partially sex-linked. Partial sex-linkage has been described for some animals but no unequivocal case is known to man.

Holandric or Y-linked inheritance: Genes which are located on the Y chromosome are said to be holandric and commonly are transmitted from fathers to sons. Several supposed instances of human Y-linked inheritance have been reported, but since they depend upon single kindreds, they are doubtful.

Figure 11a. A Kindred of Hemophilia
(after Neel and Schull, 1954)

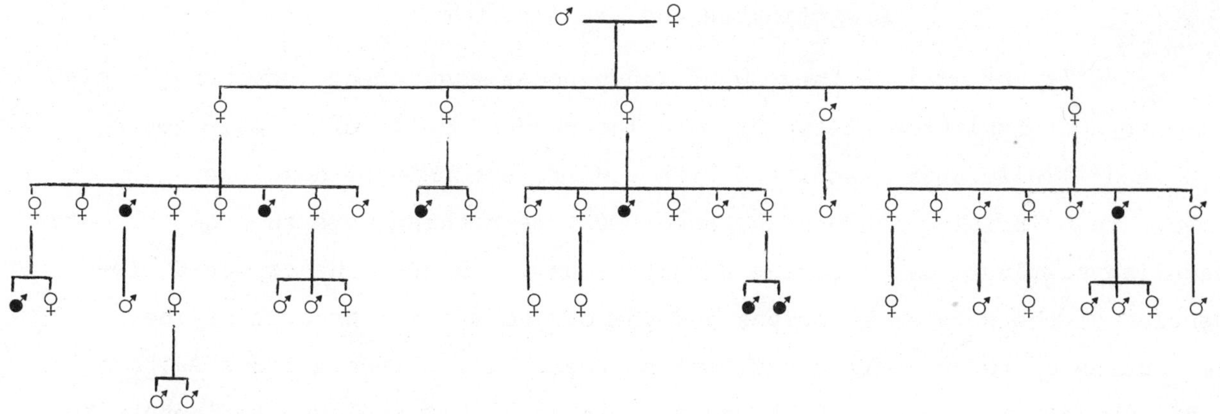

♂ = hemophilia
♀ = normal

Figure 11b. A Kindred of Hemophilia with Hemophiliac Females
(after Mersky, 1951)

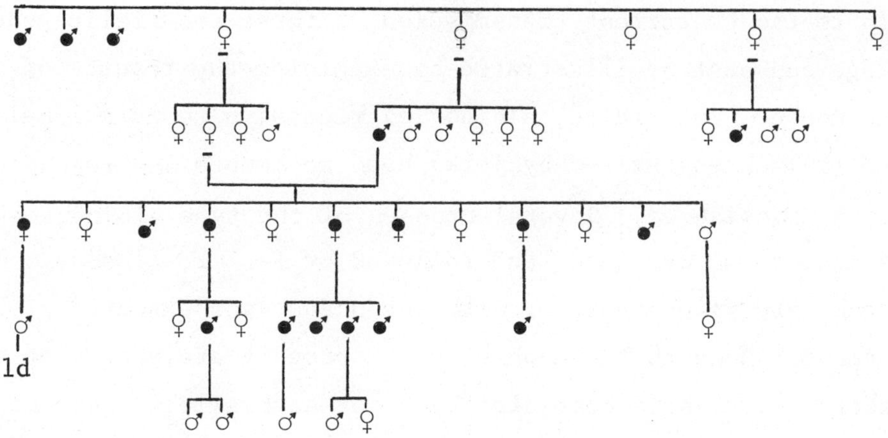

† as child

♂ = hemophiliac
♀ = heterozygote
♂ = normal

The one possible holandric trait is hairy ears which for most kindreds seems to be transmitted exclusively from fathers to sons. However, because of variable penetrance, the association of hairy ears with the Y chromosome is not absolute.

14. Linkage and Crossing Over

Mendel's second law, the law of independent assortment, governs the simultaneous transmission of two or more independent pairs of allelic genes, with each allelic pair associated with a different pair of homologous chromosomes. In principle, independent assortment is nothing more than two or more Mendelian segregations occurring simultaneously. Hence, the expected frequencies of the several genotypes and phenotypes are the product of the frequencies of two or more concurrent segregations. Given a cross $\underline{Aa;Bb}$ X $\underline{Aa;Bb}$ (where $\underline{A}$ is dominant to $\underline{a}$ and $\underline{B}$ dominant to $\underline{b}$), the expected genotypes and phenotypes among the progeny follow from the predicted segregations for each allele pair. Since for the cross $\underline{Aa}$ X $\underline{Aa}$ the expected phenotypes are $\frac{3}{4}\,\underline{A-} + \frac{1}{4}\,\underline{aa}$ and for cross $\underline{Bb}$ X $\underline{Bb}$ the expected phenotypes are $\frac{3}{4}\,\underline{B-} + \frac{1}{4}\,\underline{bb}$, their joint occurrence is $(\frac{3}{4}\,\underline{A-} + \frac{1}{4}\,\underline{aa})$ X $(\frac{3}{4}\,\underline{B-} + \frac{1}{4}\,\underline{bb})$ or $\frac{9}{16}\,\underline{A-;B-} + \frac{3}{16}\,\underline{A-;bb} + \frac{3}{16}\,\underline{aa;B-} + \frac{1}{16}\,\underline{aa;bb}$. For the testcross $\underline{Aa;Bb}$ X $\underline{aa;bb}$, the expected phenotypes among the progeny are $\frac{1}{4}\,\underline{A-;B-} + \frac{1}{4}\,\underline{A-;bb} + \frac{1}{4}\,\underline{aa;B-} + \frac{1}{4}\,\underline{aa;bb}$.

The occurrence of a number of independent traits all manifesting sex-linked inheritance raises the following question. Given two independent pairs of allelic genes whose transmission is invariably linked to the same pair of homologous chromosomes whether X chromosome or autosome. What rules of inheritance pertain to the concurrent transmission of these nonallelic genes? The rules of linkage can best be illustrated by examining the results of specific crosses in the mouse. Two traits, albino ($\underline{c}$) recessive to normal hair color ($\underline{C}$) and shaker ($\underline{sh}$) characterized by jerky head movements and recessive to normal ($\underline{Sh}$), will be considered. Several crosses of the type albino;shaker X homozygous normal;normal were made and followed by F_1 X F_1 crosses. In table 8a the F_2 progeny are given together with the phenotypes expected with independent assortment. Note that the observed F_2 progeny are wholly at variance with independent assortment. Note also that for each allelic gene pair segregation is regular: normal vs. albino = 195:66 and normal vs. shaker = 201 vs. 60. Neither segregation deviates significantly from the expected 3:1. If instead of making an F_1 X F_1 cross the F_1 individuals are testcrossed, the results are again at variance with independent assortment (cf. table 8b). Note again that segregation for each allelic gene pair is regular: normal vs. albino = 985:935 and normal vs. shaker = 976:944. Neither segregation deviates

Table 8a. Crosses of albino, shaker X normal, normal mice

$F_1 \times F_1 \rightarrow F_2$ individuals

F_2 Phenotypes	N observed	N expected*
normal; normal	192	146.7
albino; normal	9	48.9
normal; shaker	3	48.9
albino; shaker	57	16.3
total	261	260.8

Table 8b. F_1 ♀♀ X albino, shaker ♂♂ → TC individuals

TC Phenotypes	N observed	N expected*
normal; normal	941	480
albino; normal	35	480
normal; shaker	44	480
albino; shaker	900	480
total	1920	1920

*expected on basis of independent assortment.

significantly from the expected 1:1. A feature of the testcross data is that the observed parental types (normal;normal and albino;shaker) are equally frequent as are the recombinational types (normal;shaker and albino;normal). However, the parental types are far more numerous than the recombination types. How may these data be explained within the context of regular segregation? Since the results are incompatible with the independent assortment, a reasonable explanation is that albino and shaker and their normal alleles are located on the same homologous chromosomes, i. e., they are linked. How can linkage explain the results obtained? As noted earlier, many traits, in man as many as 86, manifest sex-linked inheritance. How can a large number of independent genes be accommodated on one chromosome? A simple geometric arrangement would be a linear alignment of genes along the chromosome, each gene occupying a single, non-overlapping position relative to its neighbor. This implies that so far as the length of the chromosome is concerned, some genes will be close together, others farther apart. Returning to the test cross data of table 8b, assume linkage of albino and shaker. This means the genotype of the heterozygous parent can be written as c sh/C Sh. The testcross progeny may be explained as follows. The normal;normal and albino; shaker progeny are the result of C Sh and c sh gametes formed by the heterozygous parent uniting with the c sh gamete of the albino;shaker parent. These are designated the parental types. The albino;normal and normal;shaker individuals are the result of c Sh and C sh gametes uniting with c sh gametes of the albino;shaker parent. These are called recombination types. The origin of the parental type gametes is self-evident. How are the recombination type gametes formed? It was postulated and subsequently proved that recombination involves a physical exchange of genetic material between homologous chromosomes by a process called crossing over. This means that in the heterozygous parent at some time during the first meiotic division, as a result of a crossing over event, a physical exchange occurred between a chromosome bearing C Sh and a chromosome bearing c sh resulting in gametes C sh and c Sh. To retain the integrity of the homologous chromosomes, it is assumed the crossover occurred between the two loci. As an estimate of the crossing over frequency, the quotient (number of recombination progeny)/(total progeny scored) is used. Thus, for the test cross in table 8b, the estimated crossing over frequency is (35 + 44)/1920 X 100 = 4.11%. This is sometimes designated the recombination frequency.

Another way of measuring recombination between c and sh is to first cross homozygous albino;normal (c Sh/c Sh) and homozygous normal;shaker (C sh/C sh)

NOTES

mice. The F_1 heterozygotes (c Sh/C sh) are testcrossed to homozygous albino; shaker (c sh/c sh) mice. When the progeny are scored, it is again found that the recombination progeny amount to 4% of the total. Thus, the recombination frequency (crossing over) is a function of the position of the genes on the chromosome and not how they are arranged with respect to one another on the homologous chromosomes. The term coupling (cis) linkage describes those heterozygous individuals who carry two recessive linked genes on one chromosome and the two alternative dominant alleles on the homologous chromosome, e. g., c sh/C Sh. The term repulsion (trans) linkage applies to those heterozygous individuals whose linked genes include one dominant and one recessive on one chromosome and the alternative recessive and dominant alleles on the homologous chromosome, e. g., c Sh/C sh.

Experience with a number of organisms teaches that for each pair of linked genes there is a characteristic recombination percentage. There is no objective way of estimating the recombination percentage other than by crosses with the genes involved. In general, linkage is indicated when in a test cross, the parental type progeny are significantly more frequent than the recombination type progeny.

Chromosome mapping: If linked genes are aligned linearly on the chromosome, it should be possible to order them. Consistent with the assumption that the crossing over frequency is a function of the distance between genes, it will be shown that for any three linked genes there is a unique order, i. e., one of the three must be located in the middle. This is illustrated by the results of a three gene testcross, a socalled three point cross. As demonstrated above, c and sh are linked. Other crosses demonstrate that the recessive pink eye (p) is linked to c with a recombination percentage of 20%. Since p and c are linked it follows that p and sh are also linked. What is the linear order of these three genes? Using the recessive alleles to designate each locus, three orders are possible depending upon which gene is in the middle: c-p-sh, c-sh-p, sh-c-p. Which is the correct order? From the progeny of a three point cross, the correct order can be deduced. To carry out the three point cross, a cross is first made between a homozygous, shaker, albino, pink-eyed mouse and a homozygous normal, normal, normal mouse. F_1 individuals which are perforce heterozygous at all three loci are testcrossed to homozygous shaker, albino, pink-eyed individuals and the phenotypes of the progeny scored. The cumulative results of several such test crosses are given in table 9. Before considering these results, let us consider what general results one may predict for any three point cross. If the crossing over frequency does, in fact,

NOTES

represent a measure of the distance between linked genes, then for any three linked genes whose order is a-b-c, the crossing over frequency for the interval a-b plus the crossing over frequency for the interval b-c should equal the frequency for the interval a-c. In short, as with any linear map, the distance between the two extremes must equal the sum of the intervening distances. Based upon this consideration, some generalizations can be made as to the phenotypes and frequencies expected among the three point cross progeny. Consider now the hypothetical three point cross abc/ABC X abc/abc in which the genetic distance for the interval a-b equals m, and for the interval b-c equals n. The progeny of the phenotypes ABC and abc, the non-crossover or parental types, should be the most frequent. Progeny of the phenotypes aBC and Abc, the results of a crossover between a and b of one chromosome and A and B of the homologue, taken together should occur in the frequency m, while progeny of the phenotypes abC and ABc, the results of a crossover between b and c of one chromosome and B and C of the homologue, taken together should occur in the frequency n. Finally if concurrently one crossover occurs in the interval a-b and another crossover occurs in the interval b-c, progeny of the phenotypes aBc and AbC should occur. The frequency of this last class, designated the double crossover class, should equal the product m·n, a frequency less than either m or n. By its nature, the double crossover class represents the least frequent class of progeny expected.

Turning now to the data of Table 9, it is possible to deduce a unique order for the genes involved. The progeny of the phenotypes pink-eye, shaker, albino and normal, normal, normal together make up the most frequent class and are the non-crossover types. This means that the heterozygous parent carried the recessive genes pink-eye, shaker, albino on one chromosome and the dominant alleles of normal, normal, normal on the homologous chromosomes. The progeny of the phenotypes pink-eye, shaker, normal and normal, normal, albino together constitute the least frequent class. Based upon the foregoing discussion the least frequent class arose as a result of a double crossover event or, stated another way, arose as a result of concurrent crossovers occurring on each side of the middle gene locus. Given the known parental genotype, the genes of only one locus can occur in the middle to rationally account for this double crossover class and they are the albino locus alleles c and C. Accordingly the exact genotype of the heterozygous parent may now be written as P C Sh/p c sh or Sh C P/sh c p. The remaining individuals in table 9 may be accounted for as follows: those of the phenotypes normal, shaker, normal and pink eye, normal, albino are the result of a crossover in the interval sh-c

Table 9. Three point cross involving pink eye, shaker and albino

Heterozygote (from cross pink eye, shaker, albino X normal, normal, normal) X pink eye, shaker, albino

Phenotype progeny	Number
pink eye, shaker, albino	615
normal, normal, normal	650
normal, shaker, normal	34
pink eye, normal, albino	40
normal, shaker, albino	115
pink eye, normal, normal	145
pink eye, shaker, normal	0
normal, normal, albino	3
Total	1602

crossing over %

$$\text{c-sh} = \frac{34 + 40 + 3}{1602} = 4.8\%$$

$$\text{p-c} = \frac{115 + 145 + 3}{1602} = 16.4\%$$

(or Sh-C); those of the phenotypes normal, shaker, albino and pink eye, normal, normal are the result of a crossover in the interval c-p (or C-P).

From these data it is possible to construct a simple map as follows: sh-4.8-c-16.4-p. Thus, the estimated distance between sh and p is 21.2. Note that when computing the distance (total crossing over %) for each interval, the double crossover class is included twice, once for each interval.

In principle by applying the three point cross method, it should be possible to construct a genetic map for all the chromosomes of an organism. Such a map would incorporate all the genes linked to a particular chromosome, their linear sequence and their map position. Maps of this kind have been constructed in a number of organisms (e. g., Drosophila, the mouse, corn, the tomato). It is of interest to point out that for each chromosome an estimate of its total genetic length can be computed. This is done by summing the frequencies of crossing over separating each pair of linked genes along the chromosome. When this is done genetic lengths greater than 50% and often greater than 100% are obtained. This means that in some cases linked genes may be so far apart as to show independent assortment.

15. Detection of Linkage in Man

With the exception of sex-linked traits, the establishment of linkage between genes in man is comparatively difficult. One reason for this difficulty is the large number (22) of autosomes. A second reason is that the opportunities for appropriate test cross matings of the general type AB/ab or Ab/aB X ab/ab are not frequent. In addition, the determination of genotype of the heterozygous parent may require kindred data over several generations, data which are not easy to come by. Probably the best situation for the detection of linkage employs rare dominant genes because this means that almost invariably the mating of the individual with the dominant trait is a test cross mating. In principle, linkage is sought with a socalled "marker" trait, one of such widespread occurrence that sufficient numbers of test cross marriages are bound to occur. Examples of "marker" genes are PTC tasting and the several different blood group systems. An alternative approach exploits those "recessive" traits where the heterozygote can be conclusively distinguished from both types of homozygotes. Here, too, linkage is sought with a "marker" gene. Finally, linkage is sought between different "marker" genes because the appropriate matings are comparatively easy to find.

NOTES

Direct method for detecting linkage: This method depends upon finding an almost invariable concurrent transmission of two independent traits. When such a situation is met, it is assumed that the genes involved are so tightly linked that crossing over rarely occurs to separate them. Obviously the likelihood of discovering such linkage associations is exceedingly small. Nonetheless, examples exist of which the kindreds in figure 12 are representative. The associated traits are the nail-patella syndrome (NP) and adenylate kinase (AK), an RBC enzyme. Two biochemically discrete forms of the kinase are known. They are designated as AK-1 and AK-2 and are conditioned by codominant alleles. Note that in one kindred NP is linked to AK-1 and to AK-2 in the second kindred.

A second example of linkage involves two marker genes: the Lutheran blood group system and the secretor phenotype which is associated with the secretion of certain blood group substance into the saliva and other body fluids. Since classification of everyone is possible for the alternative phenotypes (Lu^+, presence of the Lutheran antigen, vs. Lu, absence of the antigen; Se secretor or presence of blood group antigens in the saliva vs. se lack of antigens in the saliva), kindreds with appropriate test cross matings are not difficult to find. In table 10 a number of such matings are tabulated. Summing the results shows that the progeny from the coupling crosses included 28 parental types and 5 recombinational types, while for the repulsion crosses resulted in 13 parental types and 2 recombinational types. Overall 7 recombinants among 48 individuals were found, equal to a recombination value of 14%.

Indirect method of detecting linkage: The infrequent detection of linkage by the direct method led to the development of other approaches. The rationale of the socalled indirect method is straightforward. Given the progeny of the mating dihybrid and homozygous recessive, what is the probability of having obtained the observed results assuming independent assortment, i.e., no linkage is involved? If the results are not explicable by independent assortment, it is concluded that linkage is involved. An estimate of the recombination frequency is made by comparing the observed results with those expected with an arbitrarily selected range of recombination frequencies ranging from 0-50%.

Bernstein's method: This method is only of historical interest but illustrates the basic thinking involved in attempting to solve a problem. From the cross AB/ab X ab/ab progeny of the phenotypes AB, Ab, aB, and ab are

Figure 12. Linkage between the nail-patella syndrome and adenylate kinase (after Schleutermann et al. 1969)

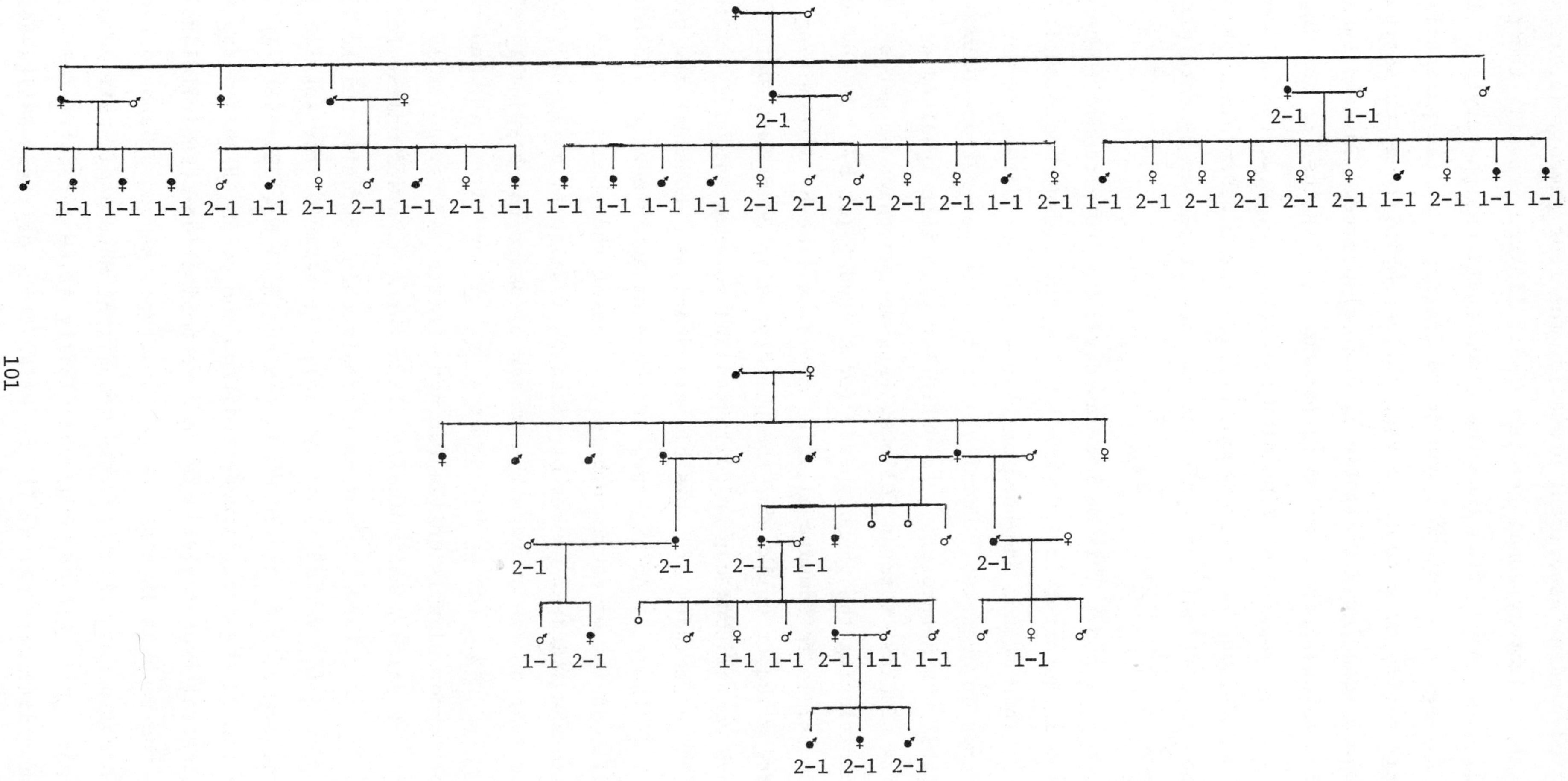

The AK genotype is indicated by the numbers below each individual: 1-1 = homozygous AK-1, 2-1 = heterozygous AK-2/AK-1

expected. If independent assortment prevails, the four phenotypic classes will occur in approximately equal frequency. If linkage prevails the parental phenotypes will occur more often than the recombinational phenotypes. The occurrence of linkage then depends upon demonstrating that the parental phenotypes are significantly more numerous than the recombinational phenotypes. For this purpose a socalled y statistic can be calculated. To make such a calculation the following information is needed: (1) whether or not the dihybrid parent is a coupling or repulsion heterozygote, a determination requiring two generations of information, and (2) the sibship size. By and large the method is difficult to apply and has been replaced by more efficient indirect methods.

Penrose's sib-pair method: This method was developed to circumvent the difficulties inherent in Bernstein's approach and the subsequent modifications thereof. To use this method one need only know that one parent is heterozygous for the genes in question. Usually data for two generations only suffice to establish a heterozygous parent. The principle involved can be described in its simplest form as follows. Assume two pairs of genes A vs. a and B vs. b conditioning the phenotypes A (or a) and B (or b). In any one sibship it is possible to compare all sibs two at a time by determining whether they are alike or unlike for the phenotypes in question. Data from a number of such comparisons can be collected and entered into a 2 X 2 table (table 11(1)) which includes all possible alternatives. Consider the cross AB/ab X ab/ab. If there is no linkage, independent assortment will occur and the sibship will consist of equal numbers of sibs of the phenotypes AB, Ab, aB and ab. With a sufficient number of sibships available, a sib-pair comparison would show that on the average the number of pairs phenotypically alike for both traits (AB vs. AB or ab vs. ab or Ab vs. Ab or aB vs. aB) and the number of pairs phenotypically unlike for both traits (AB vs. ab or Ab vs. aB) equal the number of pairs phenotypically alike for A or a and unlike for B or b (AB vs. aB or Ab vs. ab) and the number of pairs phenotypically unlike for A or a and alike for B or b (AB vs. aB or Ab vs. ab). In terms of the entries in the 2 X 2 table, the sum $n_1 + n_4$ should equal the sum $n_2 + n_3$. However, if the A and B loci are linked the progeny expected from the cross AB/ab X ab/ab should include more parental type individuals than recombination type individuals. In other words the numbers of sibs of the phenotypes AB and ab should exceed those of the phenotypes Ab and aB. This means that when a sib-pair comparison is made, more pairs will either be phenotypically alike for both traits or unlike for both traits than pairs will be alike for A or a and unlike for B

Table 10. Linkage of the Lutheran blood group with Secretor (compiled from Greenwalt 1961, Sanger and Race 1958, and Cook 1965)

Probable Genotype Parents	No. families	Phenotypes and number of children Lu^+ Se	Lu se	Lu^+ se	Lu Se	Total
$\frac{Lu^+\ Se}{Lu\ \ se} \times \frac{Lu\ se}{Lu\ se}$	10	8	20	1*	4*	33
$\frac{Lu^+\ se}{Lu\ \ Se} \times \frac{Lu\ se}{Lu\ se}$	3	2*		6	7	15

*probable recombinants

Lu^+ = Lutheran positive, Lu = Lutheran negative

Se = Secretor, se = non-secretor

or b. The same argument holds if the mating is Ab/aB X ab/ab. If independent assortment prevails, it follows that the sum $n_1 + n_4$ should equal the sum $n_2 + n_3$, but where linkage occurs the sum $n_1 + n_4$ will be significantly greater than the sum $n_2 + n_3$.

Thus the test for linkage consists simply in determining whether the observed distribution of sib-pairs among the four cells n_1, n_2, n_3 and n_4 differs significantly from that expected on the basis of independent assortment. A significant difference is interpreted to mean that linkage is involved.

In table 11 two examples are given where linkage is tested using the sib-pair method. Thus the dominant nail-patella syndrome is genetically linked to the A-B-O blood groups but not linked to the M-N blood groups.

The likelihood ratio or lod method: Pioneered by Haldane and Smith and by Morton this is now the method of choice in human linkage studies. As in the case of the sib pair method, the only information needed is the phenotypes of both parents and their children. The rationale of this method is straightforward. Given a test cross of the type Aa;Bb X aa;bb, *a priori* two genetic situations are possible. Either the genes involved are not linked and assort independently or they are linked. As already demonstrated both modes of inheritance yield identical phenotypic classes: two parental and two recombinational. However independent assortment differs from linkage in the relative frequencies of the phenotypic classes. With linkage the frequency of recombination (r) is some value less than 0.5. Hence the parental phenotypes will be more numerous than the recombinational phenotypes. With independent assortment by definition r equals 0.5. Thus the parental and recombinational classes will occur with equal frequency. These facts mean that the chance any one offspring derived from the aforementioned test cross will be of a particular phenotype depends upon whether or not the genes are linked. Assume that the genes of the heterozygous parent are linked in coupling (AB/ab) and r equals 0.1. What is the probability that the phenotype of the first child will be of the parental type (AB or ab)? Since r equals 0.1, the parental type gametes equal $(1 - r)$ or 0.9. Thus it follows that the chance the first child will be of a parental phenotype is also 0.9 or 9:1. How does this probability compare with independent assortment where r equals 0.5. Here $(1 - r)$ is 0.5 and the probability the first child will be of a parental phenotype is 0.5. Thus with linkage ($r = 0.1$) compared to no linkage ($r = 0.5$) the relative probability (odds!) that the first child will be of a parental phenotype is 0.9/0.5 or 9:5 or 1.8:1! What are the odds the first two children of the same mating will exhibit parental phenotype? When r equals 0.1 each child has a probability of 0.9. The

Table 11. Sib-pair linkage test

(1) Summation of sib-pair comparisons

		Phenotypes <u>A</u> vs. <u>a</u>: sibs alike	sibs unlike
Phenotypes <u>B</u> vs. <u>b</u>	sibs alike	n_1	n_2
	sibs unlike	n_3	n_4

with independent assortment
$n_1 + n_4 = n_2 + n_3$

with linkage
$n_1 + n_4 > n_2 + n_3$

(2) Nail-Patella syndrome and M-N blood groups†

		Syndrome: sibs alike	sibs unlike	
M-N	sibs alike	29 (30.5)*	76 (74.5)	$\chi^2 = 0.2$
	sibs unlike	35 (33.5)	80 (81.5)	p = 0.9

*( ) = random expectation

(3) Nail-Patella syndrome and A-B-O blood groups†

		Syndrome: sibs alike	sibs unlike	
A-B-O	sibs alike	78 (55.8)	44 (66.2)	$\chi^2 = 33.7$
	sibs unlike	30 (52.2)	84 (61.8)	p = <0.0001

†after Renwick and Lawler, 1954

probability of their joint occurrence is 0.9 X 0.9 or 0.81. With no linkage *r* equals 0.5 and the joint occurrence of two parental type children is 0.5 X 0.5 or 0.25. The relative probability (odds of linkage vs. no linkage) equals 0.81/0.25 or 81:25 or 3.24:1! Stated in another way, it is 3.24 times more likely that the first two children will be of a parental phenotype with linkage (*r* = 0.1) compared to no linkage (*r* = 0.5).

This simple illustration serves to emphasize a basic rule. Given a test cross: where linkage is unknown, the odds an offspring will be of a parental phenotype are invariably greater with linkage compared to no linkage. The detection of linkage by the likelihood ratio method exploits this rule. For a specific pair of traits whose linkage relationship are under study, families are sought in which a dihybrid parent is test crossed. It is assumed that coupling and repulsion heterozygotes occur with equal frequency. The progeny belonging to the parental and recombination phenotypes are enumerated. For each family two probabilities are calculated: (1) the probability the observed array of children was obtained with no linkage (*r* = 0.5) and (2) the probability the same array of children was obtained with linkage (*r* < 0.5). A relative probability (P) is calculated where

$$P = \frac{\text{probability with linkage } (r < 0.5)}{\text{probability with no linkage } (r = 0.5)}$$

Arbitrarily a P is calculated for several different linkage values e.g. 0.1, 0.2 etc. Where the value of P is greater than one, the better explanation for the array of children's phenotypes is that linkage is involved. Where P is equal to or less than one, the better explanation is that no linkage is involved.

In practice to simplify calculation the logarithm (base 10) of the probabilities is determined. The relative probability calculated as a log is called the z score or lod score (log of the odds!). For each family,

$$Z = \log \frac{p(r)}{p(r = 0.5)}$$

where *r* is some arbitrary value less than 0.5. Where more than one family of the same type is available, their combined lod score for a selected value of *r* is

$$Z = z_1 + z_2 + z_3 \ldots z_n.$$

Lod or Z scores are computed for different values of *r* ranging from 0 - 0.4. If for a specified value of *r*, Z is > 3 it is concluded that linkage is present. If Z is < -2 it is concluded that there is no linkage.

Table 12 illustrates the application of this method. Linkage between the AK locus and the two blood group loci ABO and MN is tested. The results

Table 12. Linkage analysis of the AK locus with the ABO and MN loci (after Weitkamp et al. 1969, slightly modified)

Locus	no families	Z score r = 0.05	0.10	0.20	0.30	0.40
ABO	78	+2.86	+7.95	+8.45	+5.06	+2.09
MN	62	-32.11	-17.63	-5.84	-1.48	-0.07

indicate linkage with ABO and not with MN. (It should be noted that z scores can be computed for families other than the dihybrid test cross. To simplify calculations, tables of z scores may be consulted.)

Mapping the human X chromosome: More than 80 traits in man exhibit sex linkage and while most are rare and pathologic, a few are universal and permit a mapping of the X chromosome. Because males have but one X chromosome, their X chromosome genotypes can be deduced directly from their phenotype. By exploiting specific universal sex linked traits, appropriate cis or trans heterozygous females can be directly identified from the phenotypes of their male and female parents. An estimate of the crossing over frequency between the genes involved is obtained by scoring the phenotypes of the aforementioned heterozygous females. The principle involved in this method, sometimes called the grandfather method, is outlined in figure 13a. The traits employed are (1) the Xg blood group in which the Xg^+ allele is dominant to the recessive Xg^-, and (2) the dominant $G6^+$ allele conditioning the presence of an erythrocyte enzyme, glucose-6 phosphate dehydrogenase (G6PD), deficient in $G6^-$ homozygotes. The Xg blood group is universally distributed and various ethnic groups include individuals of the two types. A deficiency of G6PD is associated with the malady, favism, afflicting individuals who eat the broad bean, *Vicia faba*, and with sensitivity to the antimalarial drug, primaquine. Thus G6PD deficiency is not uncommon in ethnic groups of the Mediterranean area and among Negroes.

NOTES

A number of kindreds have been described which exemplify the successful application of the grandfather method. Two typical examples are presented in figure 13b. The identification of parental and recombinational type grandsons is unambiguous. By extending this method to three or more sex linked traits, a unique gene sequence can be deduced as previously discussed. In this way a tentative genetic map of the X chromosome has been constructed.

16. Multiple Alleles

Thusfar, it has been assumed that for each gene locus only two alleles occur. There is now sufficient experimental evidence to support the view that a number of multiple alleles can occur at any given gene locus. A newly arisen gene mutation can be identified with and assigned to a specific gene locus if it fulfills the following criteria common to allelic genes. (1) Mutant alleles condition a common, characteristic departure from the normal phenotype. (2) Heterozygotes between independent recessive alleles display a non-complementary (non-normal) phenotype. (3) By and large during gametogenesis, alleles segregate to separate gametes. The agouti multiple allelic series in the house mouse provides an excellent example of a multiple allelic series. The locus is associated with pigment deposition of the mouse pelage. In the table below the alleles together with their associated phenotypes are seriated in the order of their dominance. Thus, $\underline{A}^y$ in compound with all other alleles results in the yellow phenotype, and $\underline{A}^w$ in compound with all subordinate alleles results in the characteristic $\underline{A}^w$ phenotype, etc. $\underline{A}$ is the wild-type allele.

	Agouti allelic series Hair color	
Allele	Back	Belly
A^y	yellow	yellow
A^w	agouti	white
A	agouti	gray
a^t	black	yellow
a	black	black

Figure 13a. Sex Linkage and Crossing over in Man

The "grandfather" method

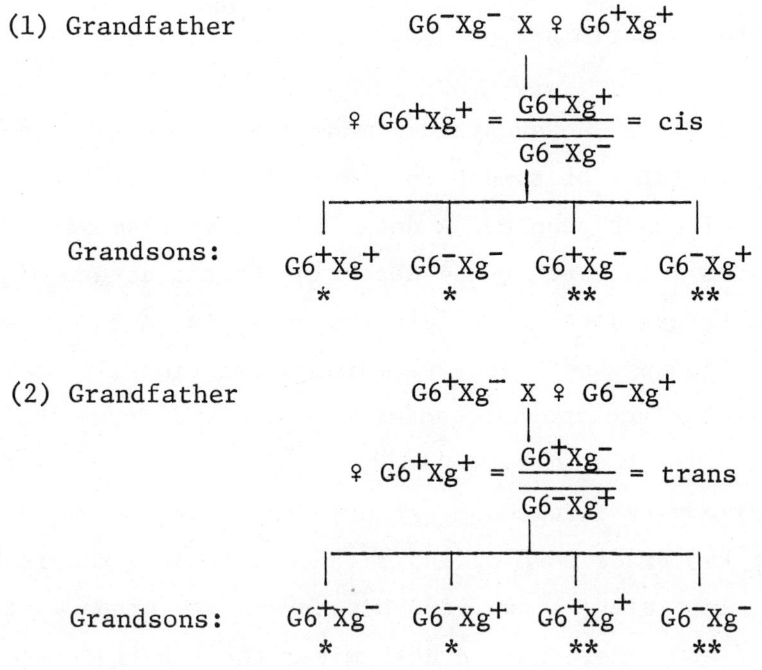

Figure 13b. Two Kindreds With X Chromosome Crossing Over

(after Stamatoyannopoulos et al. 1968)

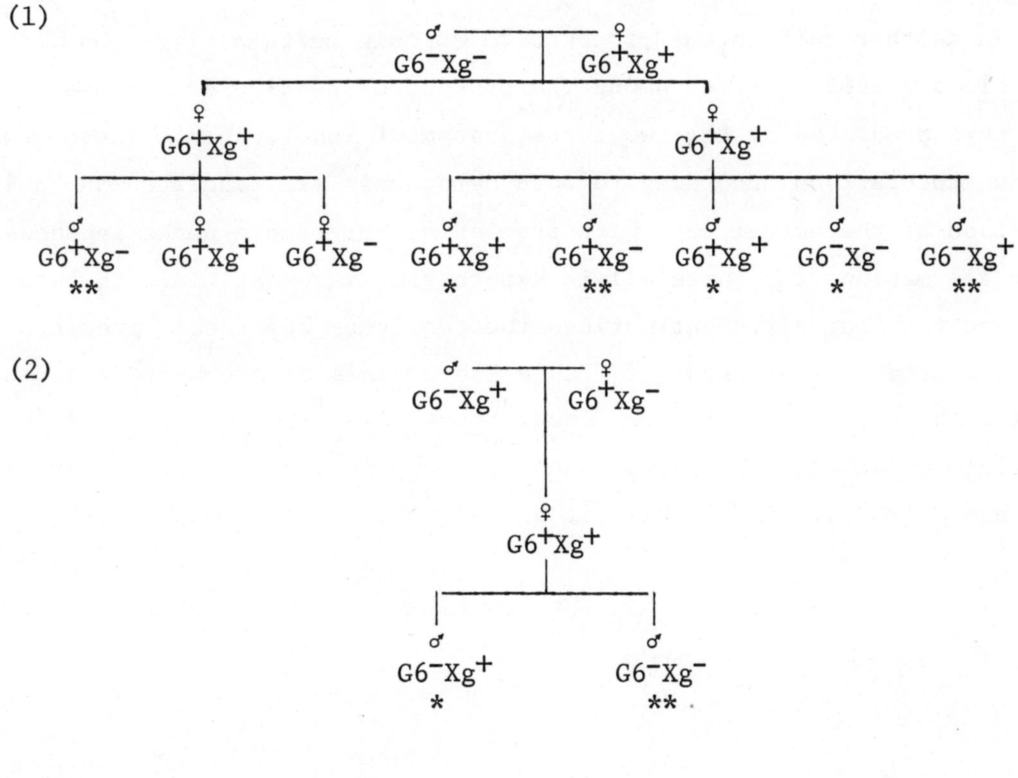

* = parental type, ** = recombination type

Seriation of alleles according to their dominance effects need not always occur. Thus, in some allelic series codominance is the rule. The number of alleles assignable to a multiple allelic series is not fixed and depends upon the acuity of the phenotypic discrimination.

The ABO blood group system: In 1900 Landsteiner made the interesting observation that the erythrocytes (RBC) of some people were agglutinated when mixed with the blood sera of other people. A detailed analysis showed that individuals could be divided into four groups depending on the nature of their RBC and sera. Landsteiner noted that only two antigens, A and B, were needed to explain the four groups and that a reciprocal relationship occurred between antigen on the RBC surface and antibodies in the blood serum. These are defined in Table 13a.

Subsequent to this discovery it was demonstrated that the ABO blood groups were inherited but the exact mode of inheritance was not established until 1924 by Bernstein. Initially, a two gene hypothesis was proposed to explain the inheritance of ABO. Here gene A determined the A antigen and the independently assorting gene B determined the B antigen. The alleles a and b when homozygous resulted in either no A or no B antigen. An alternative, three allele hypothesis in which I^a determines the A antigen, the codominant allele I^b determines the B antigen and the recessive i allele when homozygous results in neither antigen was proposed to explain certain discrepancies between blood groups observed among the progeny of particular matings and blood groups predicted by the two locus system of inheritance. The observations and expectations according to both hypothesis are tabulated in Table 11b. A comparison of the actual data with prediction for each hypothesis shows that for all matings the three allele hypothesis is compatible with observation, but for four different matings the two locus hypothesis predicts progeny not actually observed (cf. Table 13b). This discrepancy raised doubts concerning the validity of the two locus hypothesis. The adequacy of the three allele hypothesis was established by the population genetics analysis of the ABO blood groups. Since 1914, it has been recognized that racial groups

Table 13a.

The ABO blood groups defined by anti-A (α) and anti-B(β)

Blood Group	Antigens in RBC	Antibodies in Serum
A	A	β
B	B	α
AB	A,B	0
0	0	α,β

Table 13b.

ABO progeny expected according to the two locus and three allele hypothesis

Mating	Progeny phenotypes predicted from Two loci	Three alleles
O X O	O	O
O X A	O,A	O,A
O X B	O,B	O,B
A X A	A,O	A,O
A X B	A,B,AB,O	A,B,AB,O
B X B	B,O	B,O
O X AB	O*,A,B,AB*	A,B
A X AB	O*,A,B,AB	A,B,AB
B X AB	O*,A,B,AB	A,B,AB
AB X AB	O*,A,B,AB	A,B,AB

*phenotypes absent in the observed data.

differ in the frequencies of their ABO phenotypes. As one example, the U.S. Caucasian population consists of 45% O, 41% A, 10% B and 4% AB individuals in contrast to the Japanese population composed of 29% O, 42% A, 21% B and 8% AB individuals. With which genetic hypothesis are the population frequencies compatible assuming the two populations are panmictic and at equilibrium? The population expectations for the ABO blood group according to the two locus hypothesis are given in table 14. Note the relationship (A + AB) (B + AB) = AB. By substituting the appropriate observed phenotypic frequencies for the two populations into this equation, a test of the two locus hypothesis can be made by comparing computed and observed AB frequencies.

For the U.S. population, the substitution becomes:

$(0.41 + 0.04)(0.10 + 0.04)$ = frequency of AB or

0.063 or 6.3% = computed frequency of AB.

Thus, the computed frequency of AB is ca. 50% greater than the observed:

For the Japanese population, the substitution becomes:

$(0.42 + 0.08)(0.21 + 0.08)$ = frequency of AB or

0.145 or 14.5% = computed frequency of AB.

Here the computed AB frequency is almost twice the observed frequency. Thus, for both populations theory and observation disagree. The disagreement between theory and observation can be reconciled in either one of two ways: (1) either the two locus hypothesis is incorrect, or (2) neither population is panmictic. The former appears to be the more likely explanation.

The population expectations for the three allele hypothesis are given in table 15. If panmixia is assumed, it is possible to verify the hypothesis by using the several blood group frequencies in each population and substituting in the relationship $\sqrt{O + A} + \sqrt{O + B} - \sqrt{O} = 1$. The results obtained are as follows:

Population	$a = \sqrt{O + A}$	$b = \sqrt{O + B}$	$c = \sqrt{O}$	$a + b - c = 1$
U.S. Caucasian	$\sqrt{0.45 + 0.41}$	$\sqrt{0.45 + 0.10}$	$\sqrt{0.45}$	
	0.927	0.742	0.671	0.998
Japanese	$\sqrt{0.29 + 0.44}$	$\sqrt{0.29 + 0.21}$	$\sqrt{0.29}$	
	0.846	0.71	0.54	1.01

Table 14. Population analysis of ABO blood groups (Two locus hypothesis)

Let p = frequency of A $\quad$ $1-p$ = frequency of a

m = frequency of B $\quad$ $1-m$ = frequency of b

Blood Group	Genotype	Expected Frequency: Genotype	Expected Frequency: Blood Group	
O	aa bb	$(1-p)^2(1-m)^2$	$(1-p)^2(1-m)^2$	= O
A	AA bb	$p^2(1-m)^2$	$2p-p^2(1-m)^2$	= A
	Aa bb	$2p(1-p)(1-m)^2$		
B	aa BB	$(1-p)^2(m)^2$	$2m-m^2(1-p)^2$	= B
	aa Bb	$(1-p)^2\ 2m\ (1-m)$		
AB	AA BB	p^2m^2	$(2p-p^2)(2m-m^2)$	= AB
	Aa BB	$2p(1-p)\ m^2$		
	AA Bb	$p^2\ 2m\ (1-m)$		
	Aa Bb	$2p(1-p)\ 2m\ (1-m)$		

Note that $A + AB = 2p-p^2$

$B + AB = 2m-m^2$

If the two locus theory holds for the panmictic population at equilibrium,

then $(A + AB)\ (B + AB) = AB$ or

$$(2p-p^2)\ (2m-m^2) = (2p-p^2)\ (2m-m^2)$$

NOTES

Table 15. Population genetics of the A-B-O blood group system

Blood Group	Genotype	Frequency of Genotype	Frequency of Blood Group
O	ii	r^2	r^2
A	I^aI^a	p^2	$A = p^2 + 2pr$
A	I^ai	$2pr$	
B	I^bI^b	q^2	$B = q^2 + 2qr$
B	I^bi	$2qr$	
AB	I^aI^b	$2pq$	$2pq$

$p\,I^a + q\,I^b + ri = 1$

$p^2I^aI^a + 2pr\,I^ai + q^2I^bI^b + 2qr\,I^bi + 2pq\,I^aI^b + r^2ii = 1$

since $r^2 = 0$, $r = \sqrt{0}$

$0{+}A = r^2 + 2pr + p^2 = (r + p)^2$. Then $\sqrt{0{+}A} = r + p$ or $p = \sqrt{0{+}A} - \sqrt{0}$

$0{+}B = r^2 + 2qr + q^2 = (r + q)^2$. Then $\sqrt{0{+}B} = r + q$ or $q = \sqrt{0{+}B} - \sqrt{0}$

If population in Hardy-Weinberg equilibrium,

$p + q + r = 1$

or $(\sqrt{0{+}A} - \sqrt{0}) + (\sqrt{0{+}B} - \sqrt{0}) + \sqrt{0} = 1$

or $\sqrt{0{+}A} + \sqrt{0{+}B} - \sqrt{0} = 1$.

Since the findings for both populations are in complete agreement with expectation, it can be concluded that the three allele hypothesis and panmixia better explain the population frequencies of the ABO blood group phenotypes. The analysis of a variety of ethnic populations substantiate the data for the U.S. and Japanese populations.

Rhesus or Rh blood group system: In 1940 Landsteiner and Wiener reported that antibodies formed in rabbits against injected rhesus monkey erythrocytes agglutinated not only rhesus RBC but as well the RBC of 85% of the Caucasians tested in New York City. Shortly thereafter, an association was established between the socalled Rh antigens and a long known hemolytic disease of the newborn, erythroblastosis fetalis. This association will be considered in some detail below. In contrast to the ABO system where both antigens and antibodies (antisera) are normally present in blood, the Rh system is a complex of antigens whose antibodies are normally not present in human blood sera. The several Rh antigens were defined by a number of antisera derived primarily from women subsequent to pregnancy and represent an immune response. In table 16, the antigen complexes (or phenogroups) as defined by these antisera are tabulated. The genetic significance of these antigen complexes is that they are transmitted as units, i.e., they segregate to separate gametes as intact entities. Consequently, each array of three antigens is construed to be determined by a single, specific allele. For example, the common antigen complex (or phenogroup) CDe is so defined because such RBC react with three discrete antisera: anti-C, anti-D and anti-e; Cde and cde are defined in the same way. That each phenogroup is transmitted to a unit, is demonstrated in the kindred in table 16. Hence, each phenogroup is defined as the property of specific single alleles, viz., $\underline{R}_1$, $\underline{R}'$ and $\underline{r}$ respectively. Included in the table are the allele frequencies of the commonest alleles found among people of European origin.

Histocompatibility alleles: There is now quite convincing evidence that the success of tissue grafts in mammals depends on socalled histocompatibility genes. In the mouse at least 15 loci control transplantation, the most important being the H-2 locus. The success or rejection of tissue grafts depends primarily on the genotype at the H-2 locus. A number of alleles are known and allografts, or grafts exchanged between members of different strains, are successfuly only if their genotypes at the H-2 locus are identical. Where the genotypes are unlike, the grafts are rapidly rejected.

Table 16. Phenogroups and alleles of the Rh system

		Antigens (Phenogroup)							
		CDe	cDE	cde	cDe	cdE	Cde	CDE	CdE
	Allele	R_1	R_2	r	R_0	R″	R′	R_z	R_y
anti-sera	anti-C	+	-	-	-	-	+	+	+
	anti-D	+	+	-	+	-	-	+	-
	anti-E	-	+	-	-	+	-	+	+
	anti-c	-	+	+	+	+	-	-	-
	anti-d	-	-	+	-	+	+	-	+
	anti-e	+	-	+	+	-	+	-	-
	Frequency %	41	15	38	2.7	0.6	0.5	0.2	0.01

Rh phenotypes and genotypes (in parentheses)

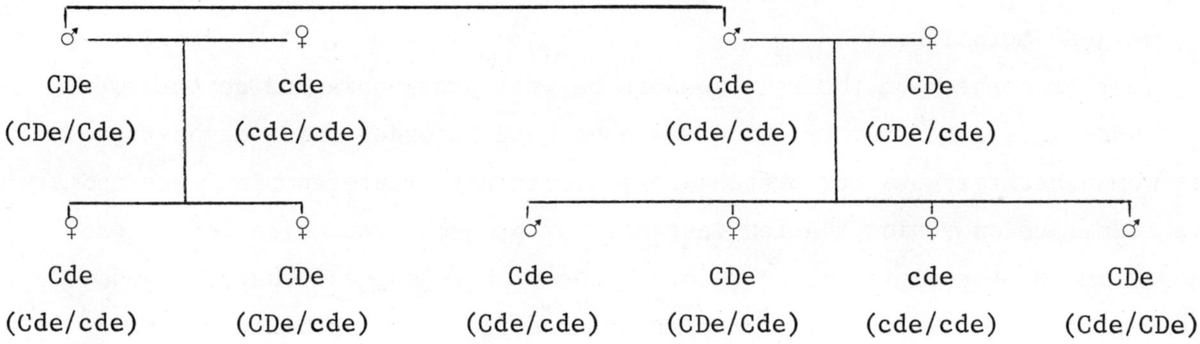

Since equivalent situations have been described for rats, guinea pigs, and hamsters, it is most likely that in man the same situation prevails.

Pseudoalleles and complex loci: Data derived from both experimental animals, plants and microorganisms are clear in demonstrating that independent mutants judged to be alleles on phenotypic and functional criteria (non-complementation of heterozygotes) do rarely segregate to the same gametes. By appropriate crosses in which marker genes were incorporated to detect the occurrence of crossing over, it was demonstrated that crossing over occurs at a low frequency ($1/10^5$-$1/10^6$) between presumed alleles resulting in gametes of two types: those containing both alleles and those containing neither allele. These exceptional types are explained as the consequence of infrequent intragenic or interallelic crossovers. Alleles which crossover are called pseudoalleles. By studying interallelic crossing over, it is possible to map the members of a multiple allelic series just as linked genes are mapped.

Interallelic complementation: Extensive studies of multiple allelic genes in several experimental organisms (e.g., Neurospora, yeast, and Drosophila) demonstrate that heterozygotes compounded of two, independent alleles do not invariably manifest a non-complementary phenotype. Rather, depending on the pair of allelic genes studied varying degrees of complementation occur. This phenomenon, designated interallelic or intragenic complementation, means that complementation is not an invariable criterion of the non-allelism of independent gene mutations.

In this connection the relationship between protanopia and deuteranopia is of some interest. Several kindreds have been recorded which demonstrate that women heterozygous for protanopia/deuteranopia or protanopia/deuteranomaly have normal color vision. A few instances of probable recombination of protanopia and deuteranopia are on record. These facts suggest that the genes involved are not allelic. The population genetics of color blindness also supports this conclusion. In the U.S. Caucasian population the frequency of males with protanopia or protanomaly is ca. 2%, with deuteranomaly or deuteranopia ca. 6%. If the genes are alleles, the overall frequency becomes 8% and the expected frequency of colorblind females should be $(8\%)^2$ or ca. 0.64%. The observed frequency of colorblind females is about 0.4%, a frequency expected if the two kinds of colorblindness involve linked, non-allelic genes. This is computed as $(2\%)^2$ plus $(6\%)^2$ or 0.04% plus 0.36% or 0.4%.

NOTES

Blood groups and disease: In 1941 Levine and associates demonstrated that the severe hemolytic disease of newborn infants (erythroblastosis fetalis) was the direct consequence of Rh blood group incompatibility between mother and child. Where precautions are not taken, erythroblastosis occurs in ca. 1/200 pregnancies in the U.S. Caucasian population, with about half the children so severely afflicted as to require a blood transfusion. Most cases of erythrobastosis involve the antigen D and the anti-D antibody with the mother dd and the child Dd. Of interest is the fact that the incidence of erythroblastosis is much lower than expected. Since in the U.S. Caucasian population dd individuals occur in a frequency of 16%, the expected incidence of erythroblastosis is estimated at ca. 9.6%. This is arrived at thus:

(1) frequency of marriage D- (DD or Dd) ♂ X dd ♀ = 84% X 16% = 13.44%

(2) frequency of Dd children from all marriages of (1) = $\frac{1}{1+q}$, where q = frequency of the d allele

$q = \sqrt{16\%}$ or 0.4; therefore, $\frac{1}{1+q} = \frac{1}{1.4} = 71.5\%$

(3) incidence of erythroblastosis = (1) X (2) = 13.44 X 71.5 = 9.6%

Note that the expected incidence of erythroblastosis (9.6%) is almost 20 times greater than the observed (0.5%). Why the discrepancy? One reason is that first born Dd children rarely if ever are affected. If the mother was "immunized," the antibody titer is too low to affect this child. A second important reason is that ABO blood transfusion incompatibility apparently protects against Rh immunization. If a dd mother is to build antibodies against D, D containing erythrocytes must cross the placental barrier, enter the maternal circulation and sensitize the mother. It was observed, as tabulated in table 17, that the likelihood of erythroblastosis is high among children whose parents are ABO transfusion compatible and low among children whose parents are ABO transfusion incompatible. Furthermore, in families where the father is AB and the mother O, no instances of erythroblastosis have been recorded. These facts suggest that where the child and mother are ABO incompatible, i.e., where the child cannot transfuse the mother, any of the child's erythrocytes containing the D antigen are rapidly removed from the circulation and thus fail to immunize the mother. Consequently, the mother and subsequent siblings are protected.

Hemolytic disease of the newborn and ABO: The association of erythroblastosis with Rh incompatibility suggests that a comparable situation involving ABO incompatibility should occur. Accordingly, one might expect A or B

Table 17.

The relationship between ABO and Rh mediated erythroblastosis fetalis

(from Race and Sanger, 1962)

Authority	Number Cases	% incidence of erythroblastosis among children of Parents ABO Compatible	Parents ABO Incompatible
Bessis	48	89.5	10.5
Levine	215	75	25.0
Race	247	84	16
Wiener	96	71	29
Broman	33	81	19
Van Loghem & Spander	240	87	13

children of O mothers to suffer from a hemolytic disease. Based on blood group frequencies, A and B children represent 12% of the progeny of O mothers. Studies show that a mild incompability does in fact occur among ca. 5% of the progeny, with ca. 0.5% showing a mild jaundice, and rarely (3 cases/1500 in one study) a transfusion is required.

The medicolegal application of blood groups: Because the blood groups phenotype of each individual is a genetically determined permanent trait which can be easily and routinely determined, they have often been employed in the adjudication of cases of disputed parentage. It should be emphasized that the blood group test is an exclusion test. This means that the test does not establish who is the parent of a child, rather it establishes who is not the parent. In most parentage litigations the dispute revolves around the putative male parent. A man can be excluded as a parent if (1) the blood group antigens he possesses are not present in the child, or (2) the alternative circumstances where the man and the mother both lack an antigen possessed by the child. The accompanying table 18 illustrates how the exclusion test works when the ABO blood group system is applied. Since it is now possible to employ seven blood group systems plus a number of serum protein systems, the precision of the exclusion test is extremely high. In fact on the basis of these tests, hardly any two individuals excepting identical twins are alike.

17. The Genetics of Sex Determination

Shortly after the rediscovery of Mendelism in 1900, it became clear with the development of microscopic techniques for the study of cell structure that each animal and plant species possessed a characteristic chromosome complement. It was further noted, especially in animals, that the chromosome complement is not identical in males and females. In males of many animal species one chromosome pair is invariably heteromorphic; this pair is designated X and Y, the sex chromosomes. In the female of the species, one of the sex chromosome pair is present twice and is designated XX. In some species, e.g., grasshoppers, the male designated XO, possesses but a single X chromosome, in contrast to the XX female. Other variations of the sex chromosome complement discovered include species where no sexual chromosome dimorphism occurs, e.g., certain amphibia, and birds and butterflies where the female

Table 18. ABO and parent exclusion

Blood group		Blood group to which
Child	Mother	father cannot belong
O	O	AB
O	A	AB
O	B	AB
A	O	O,B
B	O	O,A
B	A	O,A
A	B	O,B
AB	A	O,A
AB	B	O,B
AB	AB	O

is the heterogametic (XY) sex in contrast to homogametic (XX) males.

It soon became clear from studies with Drosophila (XX females, XY males) that the genetic determination of sexuality was not simply a matter of the sex chromosome complement. The discovery of XXY females and XO albeit sterile males discussed earlier is a case in point. Furthermore, from the cross of triploid Drosophila females by regular diploid males, a number of sex deviants were recovered including intersexes with 3 sets of autosomes and 2 X chromosomes, sterile superfemales with 2 sets of autosomes and 3 X chromosomes and sterile supermales with 3 sets of autosomes, one X and one Y chromosome. These exceptional types meant that sex could not be solely a matter of the number of X chromosomes but rather must involve a balance between sex factors on the autosomes and the X chromosomes.

The genetic basis of sex differentiation in man was essentially unclear until the etiology of individuals afflicted with the Klinefelter syndrome and the Turner syndrome was clarified. In 1949 Barr and Bertram reported that females and males could be readily discriminated on the basis of the morphology of their interphase cells. They noted that whereas in females the interphase nuclei of, e.g., epidermal cells, contained a deep staining body (Barr body), this same structure was rarely seen in males. Thus, females are said to be Barr or sex chromatin positive, males sex chromatin or Barr negative.

Individuals with the Klinefelter syndrome are phenotypic males but somewhat eunuchoid in appearance, sterile with underdeveloped genitalia and sparse body hair. The majority of Klinefelters are mentally retarded and many are institutionalized. They occur in ca. 1/400 male births. An examination of their interphase nuclei showed surprisingly most to be Barr positive. This seeming discrepancy, i.e., maleness and Barr positive, was reconciled with the finding that Klinefelters have a complement of 47 chromosomes including an XXY sex chromosome complement.

Individuals with the Turner syndrome occur in ca. 1/2500 female births. They are phenotypic females with ovarian dysgenesis, short in stature with a pronounced webbing in the neck, and usually with impaired intelligence. Turners are sex chromatin negative. The femaleness and lack of Barr body were rationalized with the finding that Turners commonly have only 45 chromosomes and are XO.

The association of Klinefelters with an XXY and Turners with an XO sex chromosome complement means that in man, in contrast to Drosophila, the Y chromosome is a strong maleness determiner. It is of interest to note that

NOTES

in the mouse the same situation occurs: XO individuals are fertile (!) females while XXY mice are males, sterile of low viability.

Subsequent to determination of the karyotypes in Klinefelters and Turners, a number of human sex chromosome variants have been described. Some of these are included in table 19.

<u>The origin of Klinefelters and Turners</u>: As noted previously non-disjunction in female Drosophila leads to the formation of XXY and XO individuals. Non-disjunction of the sex chromosomes can occur in both females and males; in fact in the mouse it was proved on genetic grounds that XO females received their X chromosome from their mothers. Therefore, the male contribution was a spermatozoon lacking both X and Y which arose following non-disjunction in the male. Red-green colorblindness provides a clue to the origin of Turners and Klinefelters. A number of colorblind Turners have been described. In at least four instances the parents of the colorblind Turner had normal vision. This suggests that the colorblindness bearing X chromosome was derived from a heterozygous mother. The Turner arose as a consequence of non-disjunction in the father resulting in XY and O sperm, the latter fertilizing an ovum with an X carrying the colorblindness gene. The accompanying kindred verifies the origin of a colorblind Turner derived from a heterozygous mother. (cf. figure 13.)

As might be expected, the incidence of colorblindness among Klinefelters is much lower than among Turners. Among the recorded colorblind Klinefelters, three are of particular interest. One such individual had a colorblind mother and a normal father. In all probability this Klinefelter arose following non-disjunction of the X chromosomes in the mother resulting in an XX ovum subsequently fertilized by a Y bearing sperm. In two other cases, the parents of the colorblind Klinefelter had normal color vision. These Klinefelters can be explained by either one of two possible events occurring in heterozygous mothers. In one event, crossing over between the X chromosome followed by non-disjunction results in an ovum with two X chromosomes each with a colorblind gene. The alternative event is non-disjunction at the second meiotic division resulting in a two X ovum each X with a colorblindness gene. Fertilization of this ovum by a Y bearing sperm results in a colorblind Klinefelter.

Recently more extensive data on the origin of Klinefelters and Turners has been obtained by use of <u>Xg</u> blood group typing. It will be recalled that <u>Xg</u> is a sex-linked blood group system where presence of the antigen ($\underline{Xg}^{+}$) is

Table 19. A Sampling of Human Sex Chromosome Variants

Sex Chromosome Complement	Number Barr Bodies	Phenotype
XXX	2	3X syndrome = ♀♀ some normal, some mental defective and sterile. ca. 1/1250 ♀ births
XXXX	3	4X syndrome = ♀ mental defective, normal gonads
XXXY	2	Klinefelter
XXXXY	3	Klinefelter, severe mental retardation
XYY	0	normal ♂
XXYY	1	Klinefelter
XXXYY	2	Klinefelter
XX	1	Sterile ♂♂ with Klinefelter phenotype

Figure 13. A kindred of a Turner female with colorblindness (after Lenz, 1957)

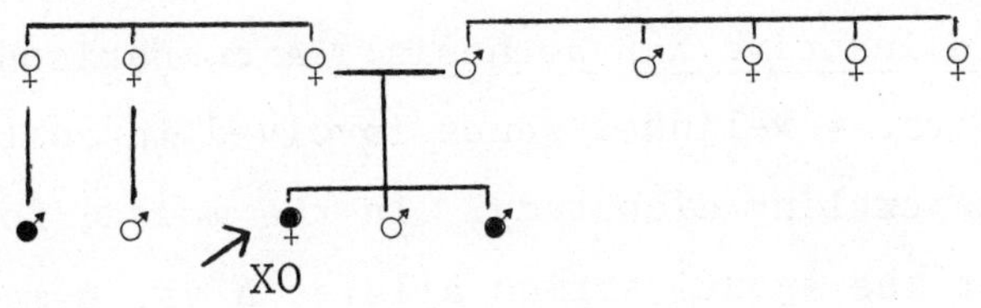

♀♂ = normal color vision

●♀●♂ = colorblind

dominant to absence of the antigen (Xg). In table 18a the Xg blood group frequencies of Klinefelters, Turners and regular males and females are given. It is apparent that while the Klinefelter frequency approximates that of females, there is nonetheless a deficiency of Xg^+ individuals. Similarly the Turner Xg^+ frequency is somewhat greater than the male frequency. Table 20b includes the Xg blood types of Klinefelters and their parents and table 20c includes Xg blood groups of Turners and their parents. The data are essentially self-explanatory and demonstrate that non-disjunction can occur in either parent to give rise to a Klinefelter or Turner offspring.

It should be emphasized that the determination of the sex chromosome complement at fertilization is not an absolute guarantee that normal differentiation of sex will occur. Instances of environmental changes during development which lead to sex transformation or intersexuality are well known. In addition, there are genetic changes which can alter the pattern of sex differentiation. An example in man of genetic alteration of sex differentiation is represented by the kindred of hereditary male pseudohermaphroditism in figure 14. The affected individuals are more correctly designated intersexes. They are XY in sex chromosome complement but resemble and regard themselves as females. They possess testes and accessory male sex organs as well as a small vagina. The kindred demonstrates intersexuality to be transmitted by normal females with the data suggesting that a dominant gene is involved with manifestation only in XY individuals, or sex linkage with manifestation only in ♂♂. Recently male pseudohermaphroditism has been described in the mouse. Genetic analysis implicates a sex linked recessive gene. By analogy the human cases are also sex-linked.

The Barr body and the inactive X hypothesis: A comparison of the phenotypic effects of autosomal vs. sex-linked genes involved in coat color pigmentation of mammals reveals a puzzling dichotomy. In the mouse, for example, an individual heterozygous for the agouti series alleles A vs. a manifests a uniform coat color phenotype indistinguishable from the agouti phenotype of the homozygous A mouse. This appears to be the general rule for other autosomal genes affecting coat color pigmentation, viz. heterozygotes manifest a uniform phenotype just as homozygotes do. However, in the case of sex-linked genes affecting coat color quite another situation is encountered. Here heterozygous females manifest a non-uniform or mosaic coat color phenotype. Some areas exhibit the phenotype of one allele, other areas the phenotype of the alternative allele with the overall distribution of areas occurring in a hit or miss fashion. Why this difference in phenotypic effects of autosomal and sex-linked genes?

Table 20a. Xg blood group frequencies in Klinefelters and Turners (from Race and Sanger, 1969)

	Xg^+	Xg	N
Klinefelters	0.825	0.175	252
Females	0.884	0.116	
Turners	0.697	0.303	261
Males	0.659	0.341	

Table 20b. Xg blood groups in Klinefelters and their parents (from Race and Sanger, 1969)

Xg blood group of			Source	Number
Father	Mother	XXY	Non-disjunction	Cases
+	+	-	mother	2
+	-	+	father	3
+	-	-	mother	5

Table 20c. Xg blood groups in Turners and their parents (from Race and Sanger, 1969)

Xg blood group of			Source	Number
Father	Mother	XO	Non-disjunction	Cases
-	+	+	father	39
+	-	+	mother	3
+	+	-	father	19
+	-	-	father	6

It has been suggested that the difference relates to properties of autosomal and X chromosomes in the somatic cells of mammals including man. Given any pair of autosomes, both homologues and the genes they contain function. In contrast given a pair of X chromosomes, only one X functions and the other X is inactive. In other words in the somatic cells of mammals the genes of one X chromosome function, those of the homologous X do not. Which X chromosome is active is a matter of chance: in some cells one X, in other cells the homologous X with equal likelihood.

Significant data supporting the X inactivation (Lyon) hypothesis have come from a variety of sources. For example, in the case of human sex-linked retinal albinism, males with the abnormal allele are without pigment in the retina. Heterozygous females have patchy pigmented retinas, i.e. some areas with pigmentation, other areas without pigmentation as though in the former only the normal allele is functional while in the latter only the abnormal allele is functional.

A parallel situation is found in the case of the enzyme glucose phosphate dehydrogenase considered earlier and conditioned by the sex-linked alleles $G6^+$ and $G6^-$. Normal enzyme activity is found in the RBC of $G6^+$ males and no enzyme activity in the RBC of $G6^-$ males. Heterozygous females have an enzyme activity intermediate between that of homozygous $G6^+$ and homozygous $G6^-$ females. Furthermore the RBC of heterozygotes fall into two classes: those with enzyme activity, those without. This implies that in some cells one X chromosome is functional, in other cells the alternative X. When epidermal cell cultures of heterozygous females are initiated from single cells, the "clones" fall into two equal classes: those with full enzyme activity, those without enzyme activity. These results are compatible with an inactive X hypothesis.

The cytological counterpart of X inactivation is the Barr body. Presumably the Barr body represents an inactive X. If as a rule only one X is active and all homologues are inactive, the predicted number of Barr bodies should be: normal ♂ = 0, normal ♀ = 1, XXX♀ = 2, XXXX♀ = 3. This prediction has been confirmed.

Figure 14. A kindred of male pseudohermaphroditism (after Petterson and Bonnier, 1937)

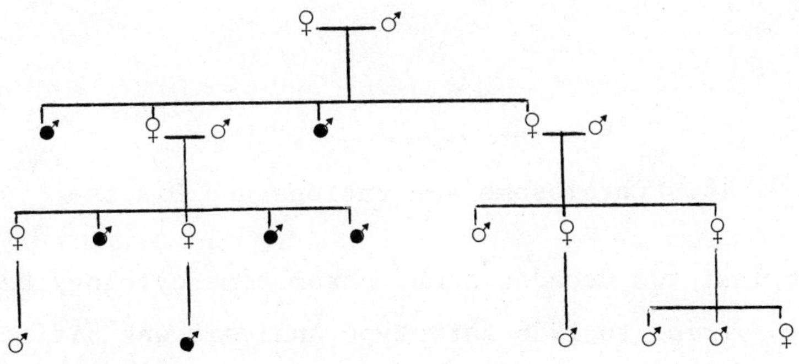

● = intersex

♂ = normal

18. Chromosome Aberrations and Disease

During the last two decades human chromosome cytology has undergone a dramatic change. Prior to 1956 karyotype analysis was difficult and uncertain due primarily to the large number of human chromosomes. Now, largely as a result of improvements in technique the analysis of human karyotypes and the identification of chromosome aberrations can be carried out with great precision in an almost routine manner. These improvements include: (1) short term *in vitro* culture of lymphocytes to yield numerous dividing cells, (2) colchicine treatment to stop mitosis at metaphase, (3) swelling the cells to disperse the chromosomes, and (4) staining procedures which facilitate the differentiation of individual chromosomes according to size, centrome location and unique morphology or banding. In figure 15a human lymphocyte metaphase chromosomes are schematically represented. In figure 15b the same chromosomes are schematically figured to illustrate their characteristic banding morphology after alkali treatment and staining by the socalled Giemsa method. Note the characteristic banding of each chromosome. The association of a number of diseases with specific chromosomal aberrations emphasizes the importance of human cytology to medicine. Selected examples follow.

Chromosome duplication: Down's syndrome, sometimes called mongolian idiocy, was described by Langdon Down and independently by Seguin in 1866. This is the commonest definable entity associated with mental deficiency and is recognizable at birth. Physical characteristics include short stature, stubby hands and full and upward slanting eyes. Individuals with Down's syndrome are prone to infections and hence have a reduced life expectancy. Two facts relating to the incidence of this syndrome have long been recognized: (1) the frequency of the syndrome is relatively high, ca. 1/600 births, and as indicated in table 19a is the same for a number of ethnic groups; (2) the empiric risk of the syndrome is highly correlated with maternal age with the incidence extremely high among mothers 40 years and older (cf. table 19b). In 1959 Lejeune and coworkers made the intriguing discovery that individuals with

Figure 15a. Idiogram of human mitotic metaphase chromosomes showing comparative size, centromere location and satellites. (courtesy of the National Foundation)

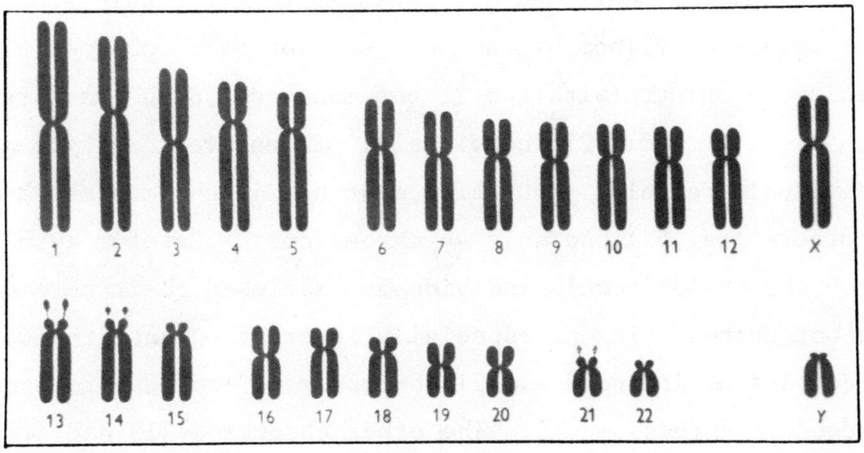

Figure 15b. Idiogram of human mitotic metaphase chromosomes showing banding after NaOH treatment and Giemsa stain. (after Drets and Shaw 1971)

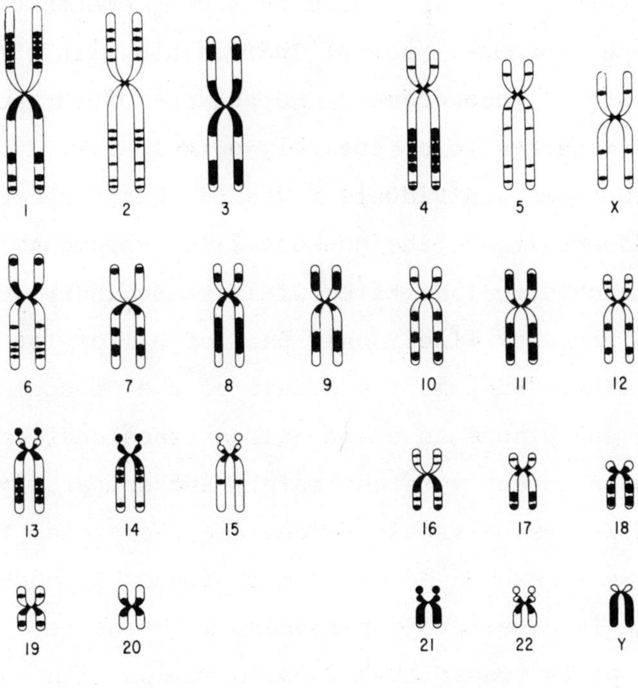

Down's syndrome possess 47 chromosomes and one chromosome, chromosome 21 is present three times (trisomic) rather than twice. (Compare the karyotypes in diagrams 3-1 and 3-2). The association of Down's syndrome with trisomy for chromosome 21 was quickly confirmed and extended to all ethnic groups. It is clear that the origin of trisomy is non-disjunction of chromosome 21 in the mothers resulting in an ovum with two 21 chromosomes. Such ova fertilized by a normal sperm yield trisomic 21 individuals. Extensive karyotype analyses of Down's individuals revealed that while most had 47 chromosomes and were trisomic 21, others (ca. 5%) had only 46 chromosomes. Careful cytological study of the 46 chromosome Down's individuals disclosed their chromosome complements were not normal. In one especially interesting case the Down's individual's complement included two 21 chromosomes, two 22 chromosomes, but only one acrocentric chromosome 15. The other chromosome 15 had its short arm replaced by a much larger arm, without doubt the long arm of chromosome 21. Thus, the two chromosomes 21 plus that part of chromosome 21 attached to 15 produce a partial trisomy (duplication) resulting in Down's syndrome (cf. diagram 3-3). A kindred analysis of the 46 chromosome Down's individual, of which the kindred in figure 16 is an excellent example, revealed that among the relatives three different chromosome complements occur: normal individuals with 46 chromosomes: normal individuals with 45 chromosomes and Down's individuals with 46 chromosomes. Karyotypes of the normal 46 chromosome showed their chromosomes to be entirely normal, i.e. 23 pairs. Karyotypes of the normal 45 chromosome individuals revealed that they possessed one chromosome 21, one chromosome 15 and the unusual 21/15 chromosome, in which the long arm of 21 is attached to 15. Effectively these individuals are disomic for both 21 and 15 since very likely only part or all of the very small arms of 21 and 15 are missing. What is the origin of the unusual chromosome? In experimental animals and plants an event called translocation or chromosome interchange occurs spontaneously rather rarely and comparatively often following irradiation with X-rays. Translocations are reciprocal events in which part of one chromosome is exchanged with part of another nonhomologous chromosome. The exchange is effectively permanent with the reciprocally exchanged chromosome parts transmitted in association with the new centromere. Translocation can account for the unusual 21/15 chromosome. If breaks in chromosomes 21 and 15 occur as indicated by the arrows in figure 16 and reciprocal exchange occurs, two new chromosomes result: 21/15 (part of 21 attached to 15 and its centromere) and 15/21 (part of 15 attached to 21 and its centromere). If the intact

Table 21a. Geographical Incidence of Down's Syndrome

Region	Frequency/Births
Chicago, U.S.A.	1/636
Honolulu, U.S.A.	1/478
Liverpool, England	1/776
London, England	1/666
Zeeland, Denmark	1/765
Australia	1/688
Hiroshima and Nagasaki	1/785
Poland	1/574

Table 21b. Maternal Age and the Incidence of Down's Syndrome

Maternal Age (years)	% Incidence
15-19	0.04
20-24	0.04
25-29	0.04
30-34	0.11
35-39	0.33
40-44	1.25
45 and >	3.15
Mean	0.15 (ca. 1/600)

NOTES

Plate 3. Representative human karyotypes. (1 = normal ♀; 2 = 47 chromosome Down's ♀; 3 = 46 chromosome Down's ♂; 4 = cri-du-chat ♀. All photos courtesy of Dr. Margery W. Shaw.)

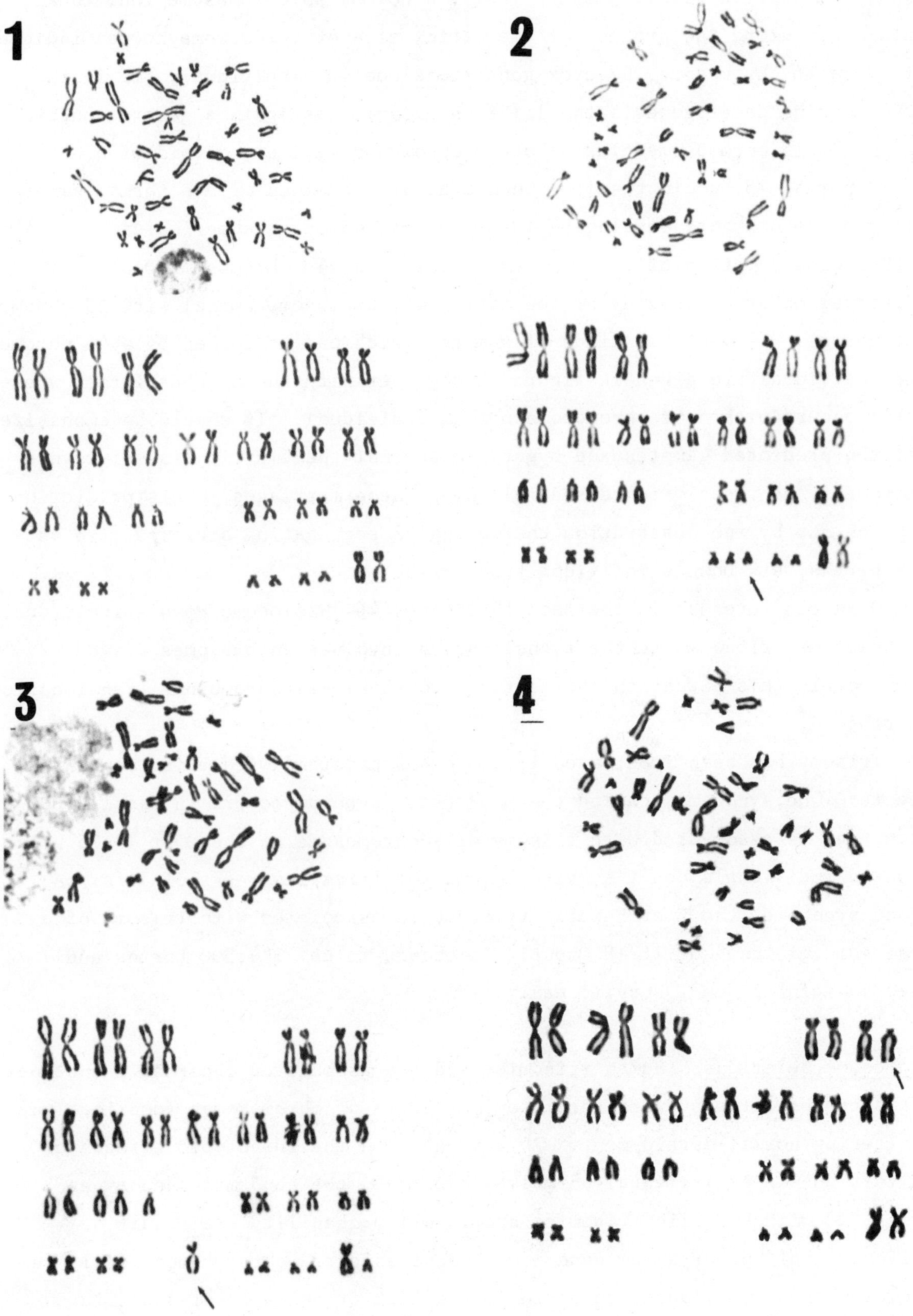

translocation (21/15 and 15/21) is transmitted to a gamete, and this gamete unites with a normal gamete, a 46 chromosome individual heterozygous for the translocation will result. Conceivably, a normal 45 chromosome individual could occur among the progeny of the mating of a 46 chromosome normal individual and a 46 chromosome, heterozygous translocation individual. If in the latter during gametogenesis the 15/21 chromosome was lost, a gamete containing the 21/15 chromosome plus 21 other chromosomes, i.e., devoid of a normal 21 and normal 15, would result. Such a gamete combined with a normal gamete will produce a zygote developing into a normal 45 chromosome individual. A further consideration of the kindred reveals that 46 chromosome Down's individual occur as progeny of the mating 46 chromosome normal with 45 chromosome normal. The several kinds of gametes which can be formed by a 45 chromosome individual are given in figure 17 together with the zygotes formed where mating occurs with a 46 chromosome normal individual. It should be emphasized that the predicted gametes and zygotes mean that the empiric risk of Down's syndrome is manifestly increased following these marriages. This prediction is borne out by the observation that among 12 segregating sibships with 48 live births, six Down's individuals occurred.

Of added interest is the fact that other 46 chromosome Down's individuals have been described where the translocation involves chromosomes 21 and 22. In principle this situation is identical to the 15-21 chromosome translocation described.

Trisomy has been implicated in two other syndrome with broad congenital defects. One syndrome, called the D or Patau syndrom occurs in ca. 1/7,600 births and is associated with trisomy of a chromosome of the D or 13-15 group. Affected individuals don't survive beyond the first few weeks of life. A second syndrome, the E or Edwards syndrome is associated with trisomy of a chromosome of the E or 16-18 group. It occurs in ca. 1/4,500 births and affected infants don't survive early life.

<u>Chromosome deletions</u>: Human trisomics and the associated abnormal phenotypes represent the effects extra chromosome or parts of chromosomes (duplication) in altering normal development. By analogy with the phenotypic effects of duplications in experimental animals and plants, the abnormal phenotypes associated with specific trisomics are in all probability the result of duplication of one or a few gene loci. Just as extra chromosomes result in a developmental imbalance, so often is the loss or deletion of specific chromosomes associated with typical, pronounced phenotypic effects. A

Figure 16. A Kindred of 46 Chromosome (T21-15) Down's Syndrome (after Shaw, 1962)

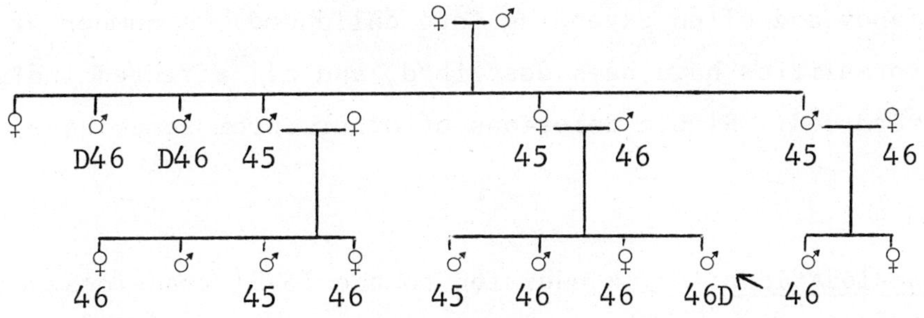

D46 = Down's syndrome and 46 chromosomes
46 = normal and 46 chromosomes
45 = normal and 45 chromosomes
↖ = propositus

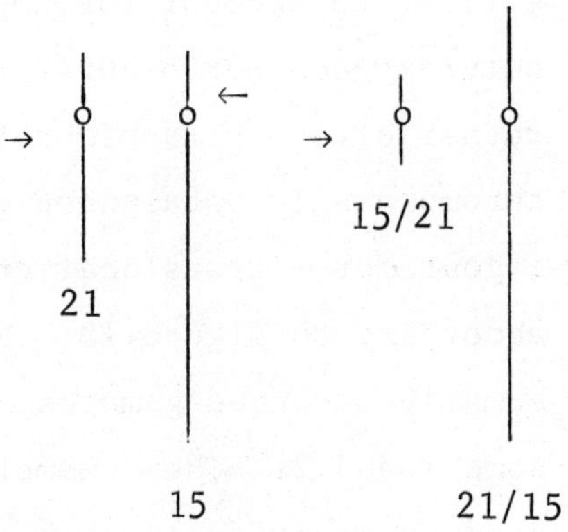

Figure 17. Chromosome disjunction in a normal 45 chromosome individual

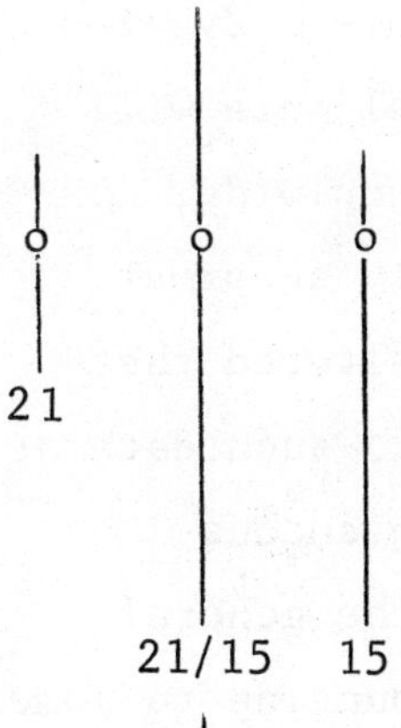

X 46 chromosome normal

	GAMETES	ZYGOTES
	21+15	46 normal
alternate	21/15	45 normal
	21+21/15	46 Down's
adjacent-1	15	45 "lethal"
	21/15+15	46 trisomic 15 = lethal
adjacent-2	21	45 lethal

striking case in point is the cri-du-chat ("cry of the cat") syndrome. To date some 100 cases have been identified and most are associated with heterozygosity for a deletion of part of the short arm of chromosome 5 (cf. diagram 3-4). The characteristic feature of the syndrome is the plaintive cry present at infancy and often extending into childhood. A number of other associated abnormalities have been described, and all affected individuals are mentally retarded. Simple deletions of other chromosomes have also been reported.

Chromosome translocations: In addition to the 15-21 translocation described earlier to account for the 46 chromosome Down's individuals, individuals heterozygous for other unrelated reciprocal translocations have been reported rather often. Translocations are of interest because they can give rise to chromosomally unbalanced gametes during meiosis. Given an individual heterozygous for a translocation, the following kinds of gametes are predicted according to figure 18. Note that the alternate disjunction leads to chromosomally balanced gametes, i.e., gametes with the intact quantity of chromosome 1 and 2. These gametes combined with gametes from a normal mate will produce individuals with a normal chromosome complement and individuals heterozygous for the reciprocal translocation in approximately equal frequency. Note further that the gametes produced by the adjacent disjunctions will be chromosomally unbalanced having an excess (hyperploid) of one chromosome and a deficiency (hypoploid) of the second chromosome. Zygotes formed by the union of these gametes with gametes from a normal mate will also be unbalanced. Depending on the degree of hyper- and hypoploidy, development of such zygotes may be only slightly or drastically abnormal. Where the imbalance is great, development may be so severely altered that either spontaneous abortions or stillbirths occur. In table 22 such data as presently available are recorded. The frequency of 21.7% spontaneous abortions is not strikingly greater than the level of 15% in the general population. However, the stillbirths plus neonatal deaths amounting to some 30% of all pregnancies is appreciably greater than the 3% for the general population and is attributable to the increased number of unbalanced gametes.

Deletion mapping: In several experimental plants and animals where the genetic and cytological conditions are favorable, deletion mapping has been successfully employed to provide an independent check on the validity of genetic maps deduced from crossing over data. Deletion mapping is feasible

Figure 18. Hypothetical Translocation Heterozygote

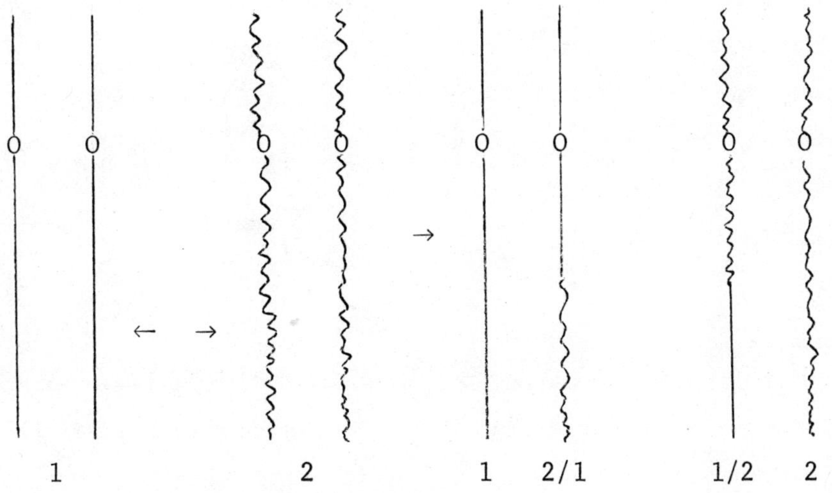

Disjunction at Metaphase Meiosis I (chromatids not drawn)

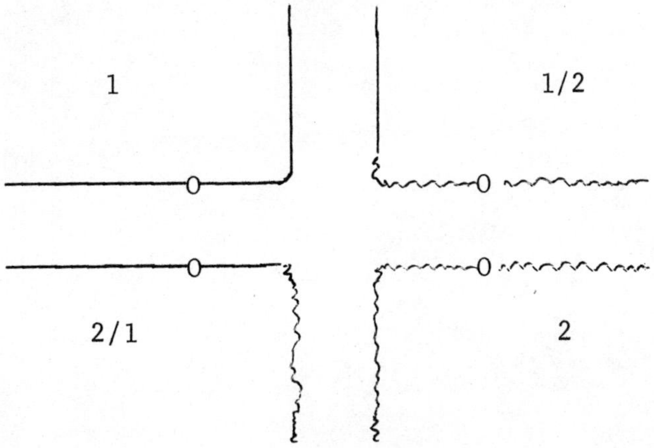

		GAMETES	
Alternate disjunction	→	1 + 2	1/2 + 2/1
Adjacent-1 disjunction	→	1 + 1/2	2/1 + 2
Adjacent-2 disjunction	→	1 + 2/1	1/2 + 2

NOTES

Table 22. Outcome of 561 pregnancies in matings of presumptive translocation heterozygotes (after Ford and Clegg, 1969)

Sex Heterozygote Parent	Number cases	Live birth: normal phenotype translocation heterozygotes	normal	not determined	Stillbirth or abnormal live birth	Spontaneous abortion
♂	55	53	30	7	50	33
♀	106	79	56	43	121	89
$\sum$	161	132	86	50	171	122

in those organisms whose chromosomes are so morphologically differentiated along their length that the loss of chromosome segments, especially small segments, is conveniently and routinely identifiable. Workable chromosomes are the meiotic midprophase (pachytene) chromosomes of a number of plant and animal species and especially the giant polytene chromosomes of many dipterous insects. The larval salivary gland chromosomes of Drosophila are unusually favorable for deletion mapping. In the salivary gland nuclei, as well as in other larval tissues, chromosome division proceeds without cytokinesis. The resulting attenuated chromosomes merge along their long axes with the homologous individual chromomeres symmetrically conjoined. What results is chromosomes of enormous size whose chromomeres take the appearance of bands of varying size throughout the chromosome length. The band pattern is characteristic and constant for each chromosome and remains constant from cell to cell, from individual to individual and from generation to generation. Finally, the homologous chromosomes join together intimately in what is called somatic pairing or synapsis. Deletions are cytologically recognized, especially in the heterozygous condition, by the absence in one chromosome of one or more bands present in the unchanged homologous chromosomes. Deletions are genetically recognized by the "uncovering" or pseudodominance of recessive genes in the heterozygous conditions. This means that where the genotype of an individual includes one chromosome carrying a specific recessive gene and an homologous chromosome in which the recessive gene locus was deleted, the individual will manifest the recessive gene phenotype. By exploiting these genetical and cytological methods it was possible to independently deletion map the X chromosome of Drosophila. The mapping procedure and results for a small chromosome segment are schematically presented in figure 19. The genetic map involves the following recessive genes: pn = prune eye color; w = white eye color; fa = facet eye structure; dm = diminutive bristles; and ec = echinus eye size. Note that the sequence deduced by crossing over (the genetic map) and the sequence deduced by considering all the deletions are identical.

The portents for deletion mapping of human chromosomes are illustrated by the two kindreds in figure 20 involving the autosomal recessive trait, pycnodysostosis, a rare bone disease. In both kindreds a cytologically discernible deletion of a G chromosome was found. In both kindreds a normal male heterozygous for a deleted G chromosome married a chromosomally normal female. In one case the female was phenotypically pycnodysostosis. The occurrence of pycnodysostosis in the offspring and in association with the deletion is interpreted as follows: the mother in one kindred was hetero-

NOTES

Figure 19. Deletion map of a portion of the X chromosome of Drosophila

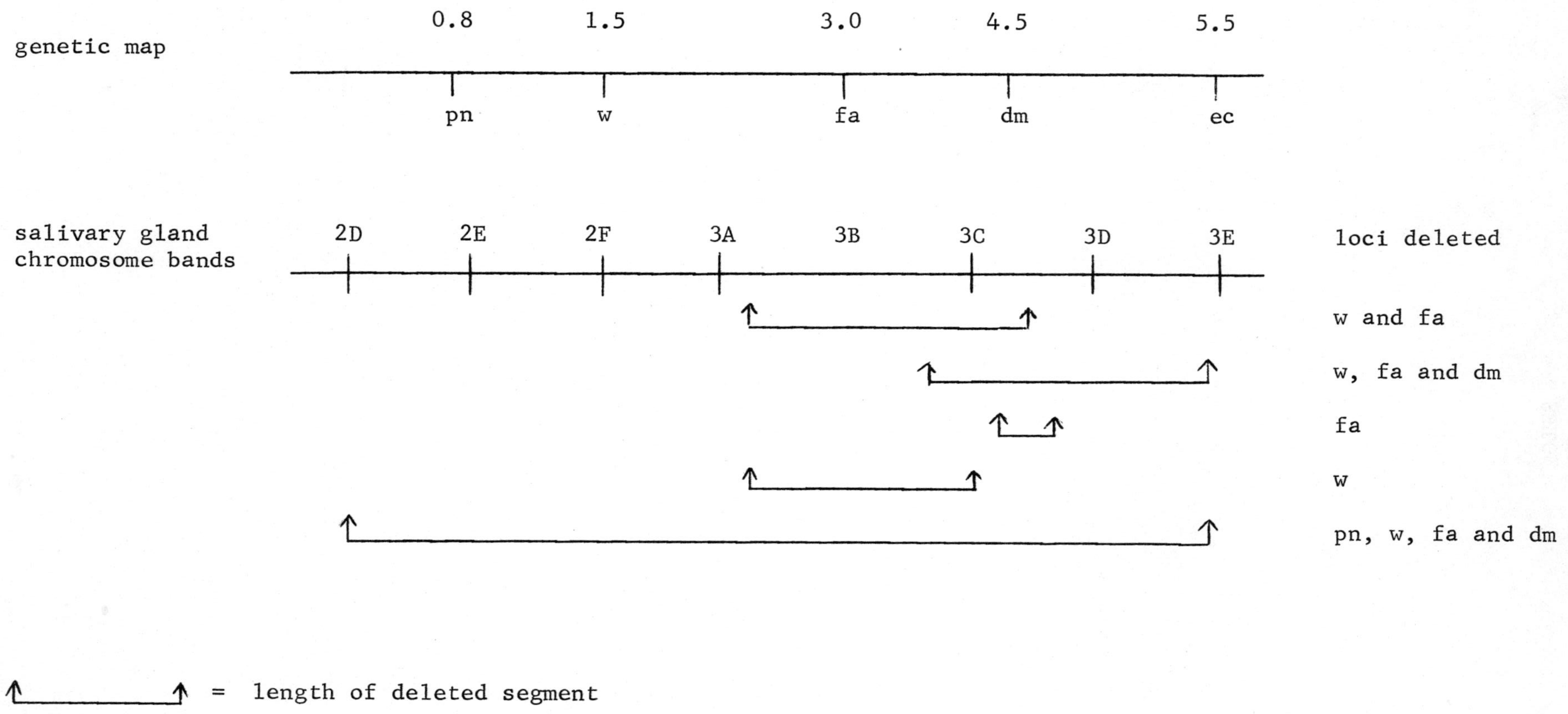

Figure 20. Deletion mapping of pycnodysostosis (after Nance and Engel, 1967)

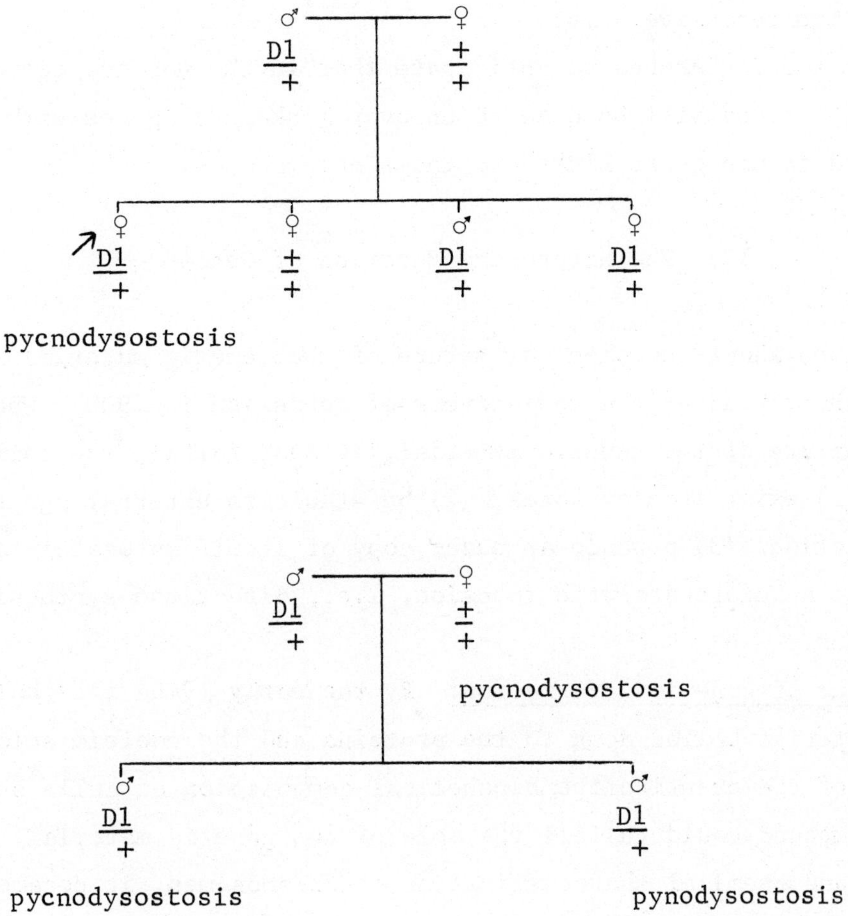

↗ = proband

$\frac{D1}{+}$ = deletion G chromosome/normal G chromosome

$\frac{+}{+}$ = normal G chromosome/normal G chromosome

zygous and in the second kindred homozygous for the recessive pycnodysostosis allele. Those offspring who manifest the trait do so because the deletion involved the loss of the pycnodysostosis locus. Thus, the combination of the recessive allele with its absence on the homologous chromosome permits the expression of the recessive allele.

It is not too farfetched to anticipate that in the not too distant future other deletions will be used to uncover linkage of genes with specific chromosomes and to map genes linked to those chromosomes.

19. The Nature and Function of Genes

Speculations about the chemical nature of the genetic material, the genes, began shortly after the rediscovery of Mendelism in 1900. Whatever the chemical nature of the genetic material, it must fulfill the following properties: (1) exist in many forms, (2) be stable to external agents and to metabolic action, (3) produce an exact copy of itself autocatalytically, and (4) possess a heterocatalytic function, i.e., direct the synthesis of numerous enzymes.

DNA as a carrier of genetic information: By the early 1940s the choice of the genetic material boiled down to the proteins and the nucleic acids. The determination of the quantitative biochemical composition of cells showed that either compound could fulfill the role of the genetic material. With the isolation and chemical characterization of chromosomes, it became clear as indicated in table 23a that two compounds are present in large amount: a basic protein (histone) and deoxyribonucleic acid (DNA). Two indirect lines of evidence favored DNA as the genetic material. (1) The action spectrum for mutations induced by irradiation with ultraviolet light corresponded closely

Table 23a. Chemical composition of chromosomes

Component	%
DNA	40
RNA	1
Basic Protein (Histone)	50
High molecular wgt protein	8

Table 23b. DNA content of various cells (g X 10^{-12}/cell)

Organism	Kidney	Liver	RBC	Sperm
Chicken	2.4	2.5	2.5	1.3
Cow	6.4	6.4		3.3
Carp		3.0	3.3	1.6
Man	5.6	5.6		2.5

to the absorption spectrum of DNA. (2) Only the DNA content of cells is highly correlated with the chromosome complement (e.g., haploid vs. diploid) as shown in table 23b. The direct demonstration of DNA as the genetic material stems from a series of experiments by Griffith in 1928 using the bacterium Diplococcus pneumoniae ("Pneumococcus"), the causative agent of mouse pneumonia. In general, two different kinds of Pneumococcus strains are recognized: (1) those which are encapsulated, form a smooth colony on agar and are virulent; (2) those which are not encapsulated, form a rough colony growth and are avirulent. In addition, each strain belongs to a particular serotype depending upon the capsule glycoprotein. Thus, one strain is designated S_I or smooth, virulent, serotype I, another is designated S_{II} or smooth, virulent, serotype II, yet another is designated R_I or rough, avirulent, serotype I, etc. In cultures it was observed that S types can mutate to R types but the reverse did not occur. Griffith made the remarkable observation that if a large number of R cells derived originally from an S_I strain were injected into a mouse, the mouse died and the surviving cells were S_I. Thus, the change of $R \rightarrow S$ which did not occur in culture took place in the mouse. To check on the source of the S_I cells, Griffith carried out two experiments. In one experiment he injected a small number of R cells together with heat killed S_I cells into mice and found that this combination not only killed the mice but that the surviving bacteria were S_I. In a second experiment he injected R cells derived by mutation from S_{II} together with heat killed S_{II} cells and again the mice died but the recovered bacteria were S_{II}. To explain the change of R cells to S_I and S_{II}, Griffith reasoned that some compound was acquired by the R cells from the killed S_I or S_{II} cells and on acquiring the compound the R cells were permanently transformed to S_I or S_{II} cells which are virulent to the mice. Subsequent experiments by other investigators established that transformation could be carried out <u>in vitro</u>. When a cell-free extract of S_{II} cells was mixed with R cells ca. $1/10^6$ S_{II} cells were recovered thereby demonstrating the existence of a transforming principle or substance.

It remained for Avery and colleagues to demonstrate in 1944 the precise nature of the transforming principle. Using the <u>in vitro</u> assay system they undertook the chemical purification and identification of the transforming principle. Their conclusion that the isolated transforming principle is purified DNA was based upon the following points: (1) the elemental compositions of the transforming principle and DNA are inseparable, (2) chemical tests revealed the isolated principle contained only DNA, (3) no immunologically active substances, e.g., proteins, could be detected, (4) treatment with the enzyme deoxyribonuclease destroyed the transforming principle, (5) the principle behaved as a highly polymerized compound as does DNA. The conclusion drawn from these findings is that the genetic information is carried in the DNA molecule. In the process of transformation exogenous DNA is incorporated into the cell's DNA thus bringing about a directed mutation and it remains as part of the cell's genetic information. Subsequent transformation experiments with Pneumococcus established that traits other than virulence could be transformed and that the transformation event does involve the physical integration of a "foreign" DNA fragment into the host cell's genome.

NOTES

RNA as a carrier of genetic information: Most plant viruses and a few animal viruses (e.g., polio and influenza viruses) are known to be composed of ribonucleic acid (RNA) and protein. This suggests that in these forms RNA is the genetic information rather than DNA. That RNA is the genetic information in these forms can be proved by an experiment involving two plant pathogenic viruses, tobacco mosaic virus (TMV) and Holmes rib grass virus (HR). Chemically each can be isolated, purified and their component RNA and protein separated. Further, the two components can be recombined, and following such a reconstitution are infective to the host plant. In this way the source of the genetic information can be determined. The basic experimental facts are outlined in table 24. Note that there is an invariable correlation between the disease symptoms (virus "phenotype") and the RNA source. This means that for these viruses RNA is the genetic information.

Structure of the nucleic acids: Nucleic acids are organized into long chains or polymers (polynucleotides) of very high molecular weight. Each chain is composed of many monomeric units or nucleotides. Chemically, a nucleotide consists of a nitrogen base, a sugar and a phosphoric acid residue. (For the chemical structure see table 25.) DNA and RNA, the two kinds of nucleic acids, differ by one nitrogen base, by their sugar residue and by a number of physical properties. The molecular composition of DNA and RNA is tabulated below.

	DNA	RNA
Purine bases	adenine	adenine
	guanine	guanine
Pyrimidine bases	cytosine	cytosine
	thymine	uracil
Sugar	deoxyribose	ribose
Acid	phosphoric	phosphoric

The structure of DNA: There is no theoretical limit to the length of a nucleotide chain and isolated DNA consists of a long chain of some 200,000 nucleotides. Chemical analysis of pure, native DNA revealed three fundamental facts. (1) Irrespective of source, the DNA contains equivalent amounts of purines and pyrimidines. (2) The amounts of adenine (A) and thymine (T) are equal as are the amounts of guanine (G) and cytosine (C),

Table 24. "Recombination" of TMV and HR protein and RNA

Strain supplying		
RNA	Protein	Disease symptoms
TMV	TMV	TMV
HR	HR	HR
TMV	HR	TMV
HR	TMV	HR

Table 25. Chemical structure of nitrogen bases, a nucleoside and a nucleotide

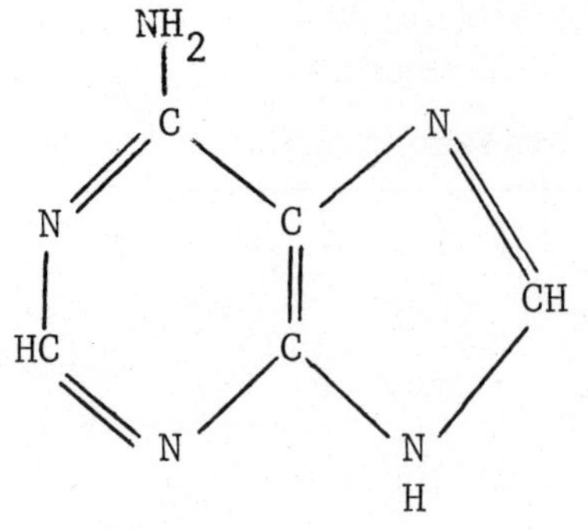

adenine

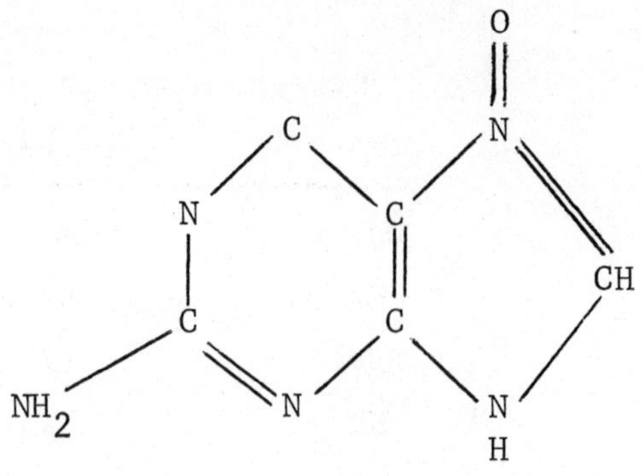

guanine

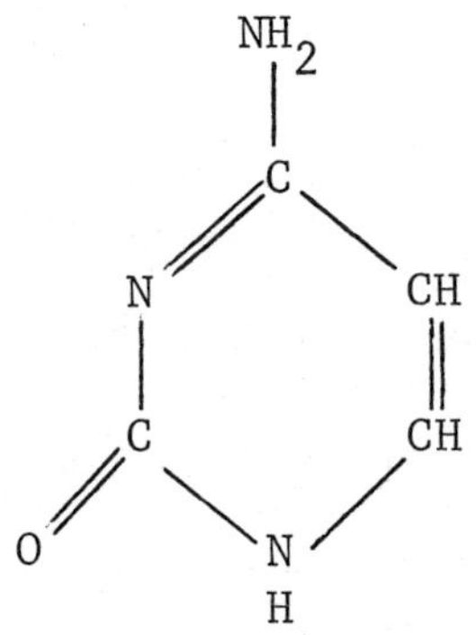

cytosine

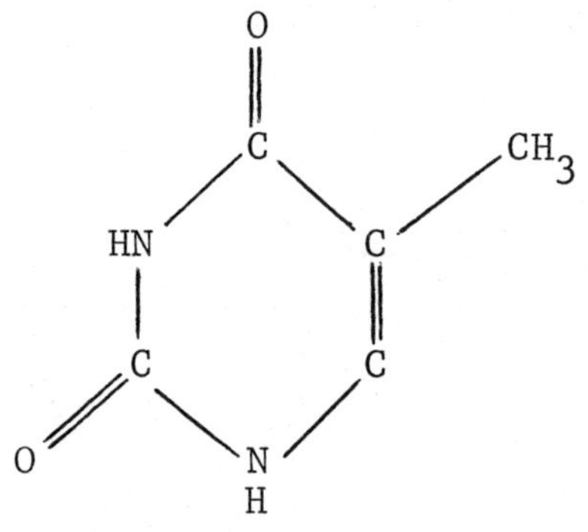

thymine

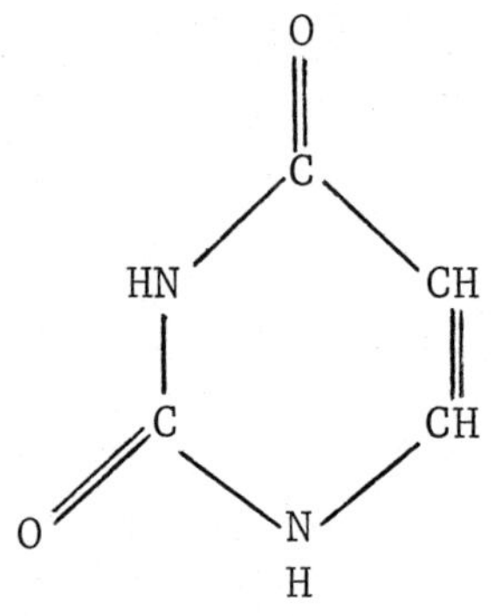

uracil

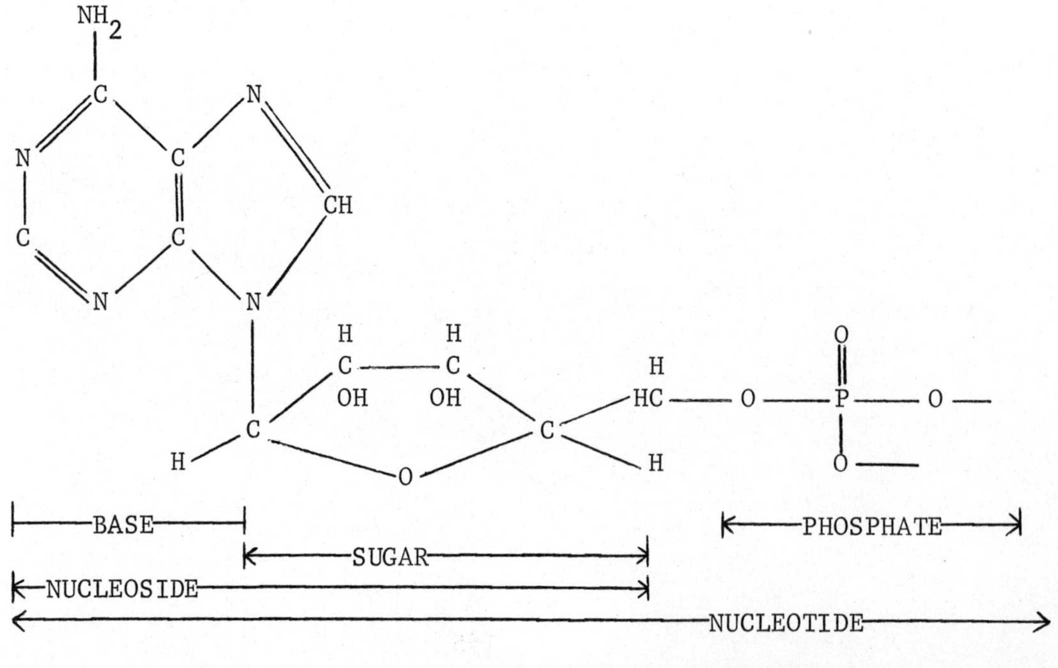

Figure 21. Double-stranded structure of DNA

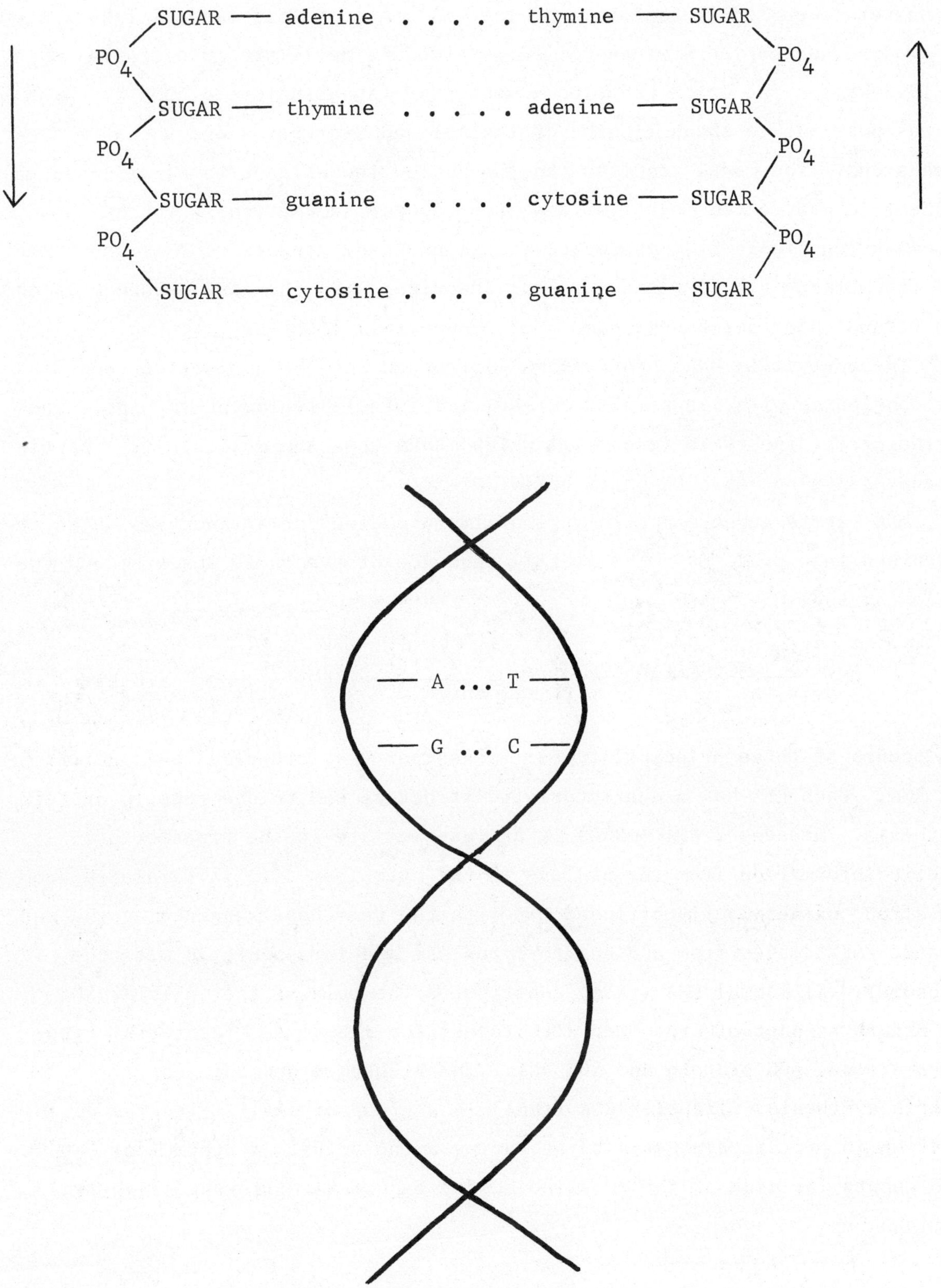

i.e., A/T = 1 and G/C = 1. (3) However, the base ratios A + T/G + C although constant within a species varies from species to species and is a species characteristic. X-ray diffraction studies, which disclose regularities in the atomic arrangement, showed that in purified DNA the same repeated pattern occurred irrespective of the DNA source. This is best interpreted as two polynucleotide strands coiled as a double helix around a central axis. In 1953 Watson and Crick proposed a model for the structure of DNA by which it was possible to account both for the A/T and G/C ratios and the atomic dimensions. The model required the two nucleotide strands to be arranged in a specific way, a double helix, and is presented schematically in the accompanying figure 21. There are two significant aspects to this structure: (1) the strands are complementary not identical, (2) the base sequence in the two strands need not be the same, but sequence is important.

In contrast to DNA, RNA is single-stranded and the nucleotides contain the four bases with, as pointed out earlier, uracil replacing thymine. The purine-pyrimidine ratio is not 1:1 which means that the molecule is a single strand equivalent to a DNA half-helix.

DNA exerts genetic control primarily in coding for amino acids which are organized into proteins. The overall sequence of events is described by the following scheme:

$$\text{DNA} \xrightarrow{\text{transcription}} \text{RNA} \xrightarrow{\text{translation}} \text{protein}$$

RNA occurs as three principal types: messenger RNA, ribosomal RNA and transfer RNA. Each RNA has a characteristic structure and unique role in protein synthesis. Messenger RNA (mRNA) is an intermediate in the transfer of genetic information from the nucleus to the cytoplasm. It is templated from one strand of the double-stranded DNA with the same base sequence as the one strand. mRNA moves from nucleus to cytoplasm and is associated with the ribosomes. Ribosomal RNA (rRNA) constitutes the bulk of the cellular RNA and occurs as part of organized structures, the ribosomes. Bacterial ribosomes are ca. 60% protein and 40% RNA. The ribosomes are the seat of protein synthesis. Transfer RNA (tRNA) is a group of small molecules which function in protein synthesis by accepting amino acids. A minimum of one tRNA occurs for each of the 20 amino acids; each tRNA differing slightly from another.

NOTES

Structure of proteins: Proteins are long-chained molecules in which various combinations of 20 different kinds of amino acids linked together. Natural proteins contain 100-1000 amino acids. The sequence of amino acids which is the linear arrangement (polypeptide chain) of the amino acids joined together by the socalled peptide linkage is a distinctive and constant feature of each protein. This is the protein's primary structure, a structure determined by the sequence of bases in the DNA. The configuration of the polypeptide backbone, or its helical configuration, refers to the secondary structure of a protein. The manner of folding of a polypeptide chain, or its topology, represents the tertiary protein structure. Finally, many proteins are aggregates of several polypeptide chains joined together in a macromolecular complex. This is a protein's quarternary structure.

The genetics of primary protein structure: Sickle cell anemia was alluded to earlier as an example of a malady conditioned by an autosomal gene where the heterozygote (sickle cell trait) can be distinguished from the two possible homozygotes - the anemic and normal individuals. Little was known about the etiology of the anemia until 1949 when Pauling and associates investigated the hemoglobin of individuals showing the three possible phenotypes, viz. those with sickle cell anemia, those with sickle cell trait and those normal. Using the method of electrophoresis which distinguishes proteins of different net electrical charge, they made the following intriguing observations. Hemoglobin of normal people behaves as though it contains one protein component, but hemoglobin of sickle cell trait individuals behaves as though it contains two components, and the hemoglobin of anemic individuals behaves as though it contains one component but different from the normal hemoglobin. One could mimic the sickle cell trait hemoglobin by mixing the hemoglobins of normal and anemic persons. The electrophoretic patterns are schematically presented in figure 22. Pauling et al. pointed out that these results could be interpreted by assuming that genes had a direct effect on the synthesis of the hemoglobin molecule with normal allele "producing" normal hemoglobin and the sickle cell allele "producing" abnormal hemoglobin. In short it was postulated that a direct relationship exists between genes and the kind of molecules produced. What now remained was to demonstrate what aspect of hemoglobin structure distinguishes normal hemoglobin (HbA) from sickle cell hemoglobin (HbS).

Hemoglobin is a complex molecule which contains four socalled heme groups each of which is joined to a separate polypeptide chain. Each heme

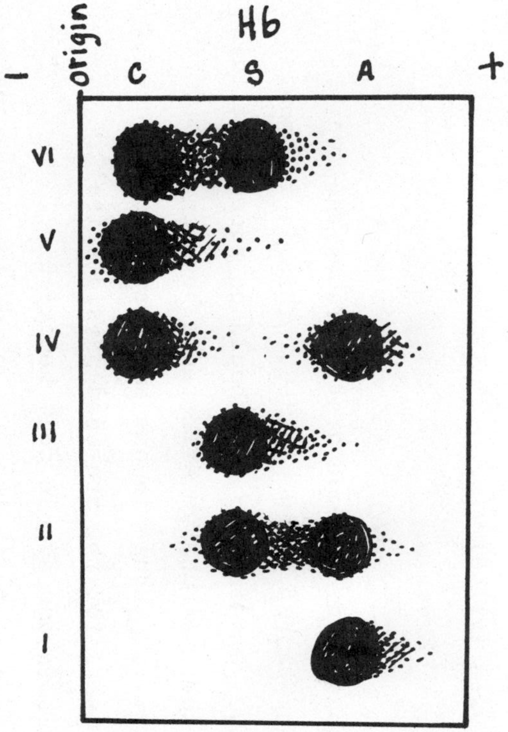

Figure 22a. Separation of human hemoglobins by electrophoresis. I = normal (HbA), II = Sickle cell trait (HbA/HbS), III = Sickle cell disease (HbS), IV = HbC heterozygote, V = HbC homozygote, VI = heterozygote HbC/HbS.

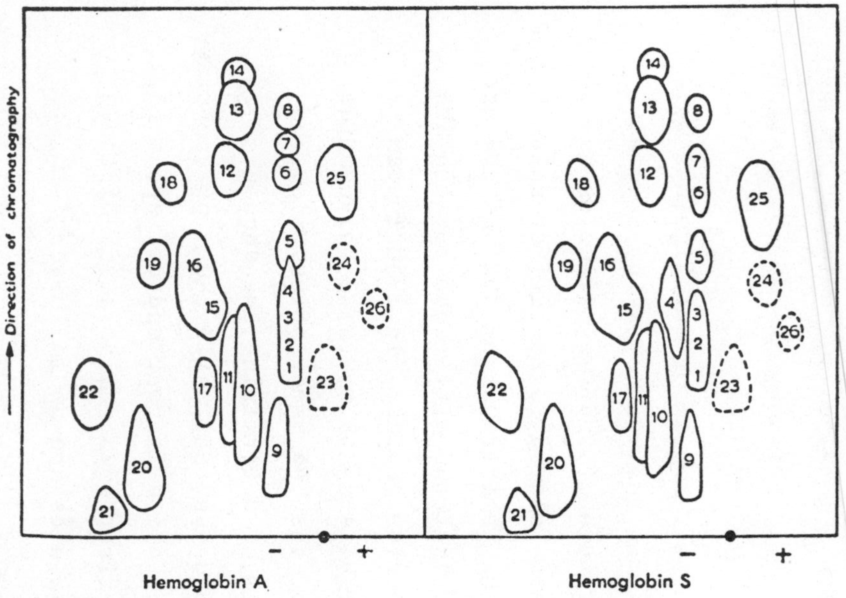

Figure 22b. "Fingerprints" of trypsin digests of Hemoglobins A and S, (after Ingram 1958) illustrating the numbering of peptide pairs.

group contains a Fe^{++} ion at its center where the oxygen attaches. The two kinds of polypeptides are designated α with a total of 141 amino acids and β with 146 amino acids. As regards these polypeptides, HbA may be designated $\alpha_2\beta_2$.

Subsequent to the study of Pauling et al. two important findings were made. HbA and HbS are indistinguishable when analyzed by X-ray crystallography. This means that the gross structure of the two molecules is identical and the difference between them is not related to any differences in the folding of the molecules. In addition, the heme groups of HbA and HbS are indistinguishable. These facts suggested that the difference between HbA and HbS more likely resides in the polypeptide chains. In 1958 Ingram was able to pin-point the difference between HbA and HbS to the β polypeptide chain. He "fingerprinted" the proteins of the two hemoglobins, i.e., he hydrolyzed them enzymatically with the aim of determining the sequences of their component amino acids. What he found was that while the α polypeptides of HbA and HbS are identical, the β chains are different and different in a consistent way. If the sequences of the first eight amino acids of the β chains of HbA and HbS are compared, the results are as follows:

	1	2	3	4	5	6	7	8
HbA	val-	his-	leu-	thr-	pro-	glu-	glu-	lys-
HbS	val-	his-	leu-	thr-	pro-	val-	glu-	lys-

(val = valine, his = histidine, leu = leucine, thr = threonine, pro = proline, glu = glutamic acid, lys = lysine)

Note that one and only one amino acid difference distinguishes the two: amino acid 6 is glutamic acid in HbA but valine in HbS. This is the only difference which separates the two hemoglobins. It should be added that this difference also explains the electrical charge difference between the two molecules. Thus, a seemingly trivial change in one amino acid conditioned by a single allelic difference makes for the profound physiological difference between normal hemoglobin and sickle cell hemoglobin.

The observations of Pauling et al. stimulated a search for additional hemoglobin variants. One such discovery was made by Itano and Neel in a family where the sickle cell trait also occurred. The relevant data are presented in figure 23. In this kindred, individual I-2 presented an

Figure 23.

A kindred with hemoglobin C

(after Itano and Neel, 1950)

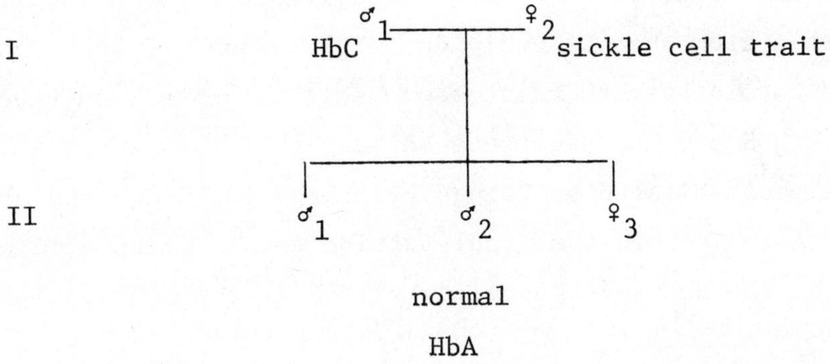

Hb of ♂ I-1, cf. figure 22a, IV

♂ II-1, cf. figure 22a, II

♀ II-3, cf. figure 22a, VI

electrophoretic pattern typical of the sickle cell trait. Her mate, I-1, also possessed two hemoglobin components, HbA and a new component electrophoretically distinguishable from HbS and designated HbC. The three phenotypes detected among their progeny include a fully normal son, a son with the sickle cell trait, and a daughter with the two abnormal components HbS and HbC. Subsequent studies uncovered individuals whose hemoglobin consisted exclusively of HbC and whose parents were heterozygotes, i.e., exhibited a phenotype inseparable from that of individual I-1. Such HbC homozygotes suffer from a mild anemia but otherwise seem healthy. Data from several kindreds where both HbS and HbC segregate suggest that they are alleles. This is conclusively established by "fingerprinting" the HbC hemoglobin. HbC differs from HbA and HbS in the β polypeptide and at amino acid number 6: the glutamic acid of the HbA and the valine of HbS is replaced by lysine in HbC.

Intensive, world-wide studies of human hemoglobins has led to the discovery of over 100 variants in the primary structure of the α and β hemoglobin chains. Some of these are without physiological effect, others result in profound anemias. A small sample of these variants and the associated amino acid changes is tabulated in table 26. The relevance of these variants to the genetic code will be considered shortly.

20. The Genetic Code

It is not unreasonable to deduce from the many Hb variants that the information specifying the precise sequence of amino acids in polypeptides is stored in the DNA. This information is transferred from DNA to mRNA which carries the information to the cytoplasm where amino acid synthesis and the conjoining of amino acids to make polypeptides takes place. In the chemical structure of DNA, one feature occurs which is the obvious candidate for the mechanism coding for specific amino acids. This is the sequence of bases in the DNA. Since this sequence is very long there are numerous possible sequences which could direct the synthesis of different amino acids. It was generally assumed, and now clearly proved, that the genetic information specifying a sequence of amino acids is stored in a colinear way in the corresponding DNA, i.e., the order of the amino acids in a polypeptide reflects the order of the coding units in the DNA.

Table 26. Amino acid substitutions in Hb A

α Chain substitutions

position	1	2	16	30	57	58	68	116	141
	val	leu	lys	glu	gly	his	asn	glu	arg
Hb variant									
I			glu						
G Honolulu				gln					
Norfolk					asp				
M Boston						tyr			
G Philadelphia							lys		
O Indonesia								lys	

β Chain substitutions

position	1	2	3	6	7	27	43	63	67	121	146
	val	his	leu	glu	glu	glu	glu	his	val	glu	his
Hb variant											
S				val							
C				lys							
G San Jose					gly						
E						lys					
G Galveston							ala				
M Saskatoon								tyr			
Zürich								arg			
M Milwaukee-1									glu		
D Punjab										gln	
O Arabia										lys	

Natural proteins contain various amounts of 20 amino acids. With four bases A, G, C and T (or U) present in DNA and RNA, it is obvious that more than one base is needed to code for an amino acid. It is also clear that the 16 combinations possible when the four bases are considered two at a time (e.g., AA, AG, AC, etc.) is also insufficient to code for the necessary 20 amino acids. If the available four bases are considered three at a time (e.g., AAA, ACA, AAC, etc.), 64 combinations are possible which is more than enough to account for the 20 amino acids.

A partial solution of the coding problem was achieved through studies with mRNA. It was possible to obtain _in vitro_ polypeptide synthesis by employing a cell-free system containing a combination of amino acids, activating enzymes, ribosomes, tRNA mixtures, an energy source in ATP and synthetic mRNA in place of natural mRNA. Moreover, it was found that the kind of polypeptide made in this _in vitro_ system depends upon the synthetic mRNA included. If, for example, the synthetic mRNA is polyuridylic acid, which corresponds to a triplet UUU, a polypeptide is produced consisting exclusively of phenylalanine molecules linked together. If, however, polyadenylic acid (AAA) is the mRNA, phenylalanine is not incorporated into the synthetic polypeptide; instead lysine is the incorporated amino acid. Subsequently, synthetic mRNA was synthesized containing two bases alternating in sequence, e.g., CUCUCU. Such a mRNA represents two triplets: CUC and UCU. It was found that the synthetic polypeptides coded by this system contained alternately leucine (CUC) and serine (UCU). By synthesizing and testing a variety of mRNA and by socalled binding studies of trinucleotides with amino acids, it was possible to assign codons (triplet base sequences) for each of the 20 amino acids. These codons, which collectively represent the genetic code, are listed in table 27. Two generalizations are possible from this tabulation: (1) the code is degenerate, i.e., a particular amino acid may be coded for by several codons, and (2) some codons, e.g., UAG, called nonsense or terminating codons, do not specifiy a particular amino acid and thus stop polypeptide synthesis. Other evidence suggests that the genetic code is nonoverlapping, commaless and quasi-universal.

Does the genetic code as deduced from _in vitro_ experiments represent the _in vivo_ code? Some insights into the reality of the code can be gained by examining the amino acid changes associated with each of the many Hb mutants uncovered. The simplest way to account for the origin of these mutant hemoglobins is to assume that each arose as a mutation, and the simplest kind of mutation imaginable is a change in one of the three bases constituting the

Table 27. GENETIC CODE m-RNA

First Letter	Second Letter: U	C	A	G	Third Letter
U	phe	ser	tyr	cys	U
	phe	ser	tyr	cys	C
	leu	ser	Term	Term	A
	leu	ser	Term	try	G
C	leu	pro	his	arg	U
	leu	pro	his	arg	C
	leu	pro	gln	arg	A
	leu	pro	gln	arg	G
A	ile	thr	asn	ser	U
	ile	thr	asn	ser	C
	ile	thr	lys	arg	A
	met	thr	lys	arg	G
G	val	ala	asp	gly	U
	val	ala	asp	gly	C
	val	ala	glu	gly	A
	val	ala	glu	gly	G

Term = polypeptide chain termination

codon for a specific amino acid. If these assumptions are correct, it means that in general each Hb mutation involves a change in only a single amino acid. Verification of the genetic code follows if the amino acid change is compatible with a change in but one base of the codon specifying the amino acid. In table 28 several examples of Hb mutations are given together with the described amino acid change and the postulated change in the codon. Note that in each case the amino acid alteration is fully compatible with a change in one base of the codon. Examples of two different mutations of the same amino acid are equally compatible with one base change. The code as given is real.

21. "Inborne Errors in Metabolism"

In 1902, the English physician Garrod published a study entitled "The incidence of alkaptonuria, a study in chemical individuality." In this paper, Garrod anticipated the relationship between genes and metabolism, and in fundamental terms the relationship between genes and enzymes (proteins). It will be recalled that alkaptonuria is a condition in which a compound, homogentisic acid, is excreted in large amounts (several grams daily) in the urine. Homogentisic acid oxidizes rapidly on exposure to air to a black pigment, alkapton. Garrod developed a number of basic facts relevant to this condition: (1) people are either alkaptonurics or normal, (2) alkaptonuria occurs more frequently in certain families and especially among the offspring of consanguineous marriages, (3) alkaptonuria is best explained as determined by an autosomal recessive allele, (4) the excretion of excessive amounts of homogentisic acid in alkaptonuria is postulated to be associated with a defect in the metabolism of the amino acid tyrosine. In addition, Garrod suggested that genetically controlled metabolic defects are not unique but of common occurrence. By 1908 he had described a number of these including cystinuria, porphyria and pentosuria. These he collectively called "inborne errors in metabolism." To explain the inborne errors, Garrod postulated that in each case the body is no longer able to perform some particular step in normal metabolism presumably because of the absence or non-functioning of a specific enzyme. Thus, he suggested a gene-enzyme relationship.

As is so often the case, Garrod's ideas were far ahead of their time and much of what he did was either ignored or not understood by his contemporaries. It was not until some thirty years later following the experiments of

Table 28. Verification of genetic code from representative hemoglobin mutants

	Hb Type	Amino Acid Change	One-Step Code Letter Change	Base Pair Change
β6	S	glu → val	GAG → GUG	A → U
	C	glu → lys	GAA → AAA GAG → AAG	G → A
β7	G (San Jose)	glu → gly	GAA → CAA GAG → CAG	G → C
β67	M (Milwaukee)	val → glu	GUG → GAG	U → A
α16	I	lys → glu	AAA → CAA AAG → CAG	A → C
α58	M (Boston)	his → tyr	CAU → UAU CAC → UAC	C → U
β121	D (Punjab)	glu → gln	GAA → CAA GAG → CAG	G → C
	O (Arabia)	glu → lys	GAA → AAA GAG → AGA	G → A
β63	Zürich	his → arg	CAU → CGU CAA → CGC	A → G
	M Saskatoon	his → tyr	CAU → UAU CAC → UAC	C → U

Beadle and Tatum with the bread mold, Neurospora, that Garrod's ideas were understood, appreciated and convincingly established. The rationale of experiments by Beadle and Tatum which led to the concept of "one gene-one enzyme" was, in retrospect, simple. Neurospora is chemically a comparatively complex organism containing proteins, carbohydrates, nucleic acids, lipids, etc. Notwithstanding, a fully differentiated normal organism can develop from a single cell growing on a simple (minimal) medium containing a number of inorganic salts, a carbon source (e.g., sugar), a nitrogen source (e.g., nitrate) and biotin, a B vitamin. How are differentiation and growth of the fully developed organism accomplished? In the simplest terms Neurospora must be capable of synthesizing from the nutrients included in the minimal medium the variety of complex organic molecules (proteins, nucleic acids, etc.) necessary for growth and differentiation. The various syntheses must be enzymatically controlled; and the synthesis of the necessary enzymes is mediated by various Neurospora genes. If these arguments are valid, and especially if genes are coding for specific enzymes, it should be possible to mutate these genes such that they now code for mutant enzymes no longer able to catalyze specific syntheses. These mutant Neurospora should now have nutritional requirements not reflected by the unmutated (wild-type) strains.

The method devised for isolating socalled nutritional mutants in Neurospora is outlined in figure 24. In principle wild-type, haploid, vegitative cells or conidia are irradiated with X-rays or UV, agents known to increase the mutation frequency, and are crossed to an unirradiated strain. Subsequent to meiosis the haploid ascospores, the descendants of both the irradiated and unirradiated conidia, are isolated individually on a complete medium. (The complete medium contains an array of complex molecules synthesized by the wild-type Neurospora.) Conidia derived from the isolated single ascospores are tested for their ability to grow on the minimal medium. If growth occurs, the isolate is wild-type, i.e., no mutation was induced; if no growth occurs, the isolate is presumed to be mutant, i.e., carries a mutation which no longer permits the completion of some biosynthetic step. Subsequent genetic and biochemical studies are carried out to identify and characterize each mutant recovered. With these procedures it was possible to isolate literally hundreds of mutants each incapable of growing on minimal medium but capable of growth when minimal medium is supplemented with a single specific compound, e.g., a single amino acid or a B vitamin, etc.

Detailed study of a number of independent mutants satisfied by the same nutritional supplement revealed that all are not identical (alleles) and many

Figure 24. Outline procedure for detecting biochemical mutants in Neuropora.

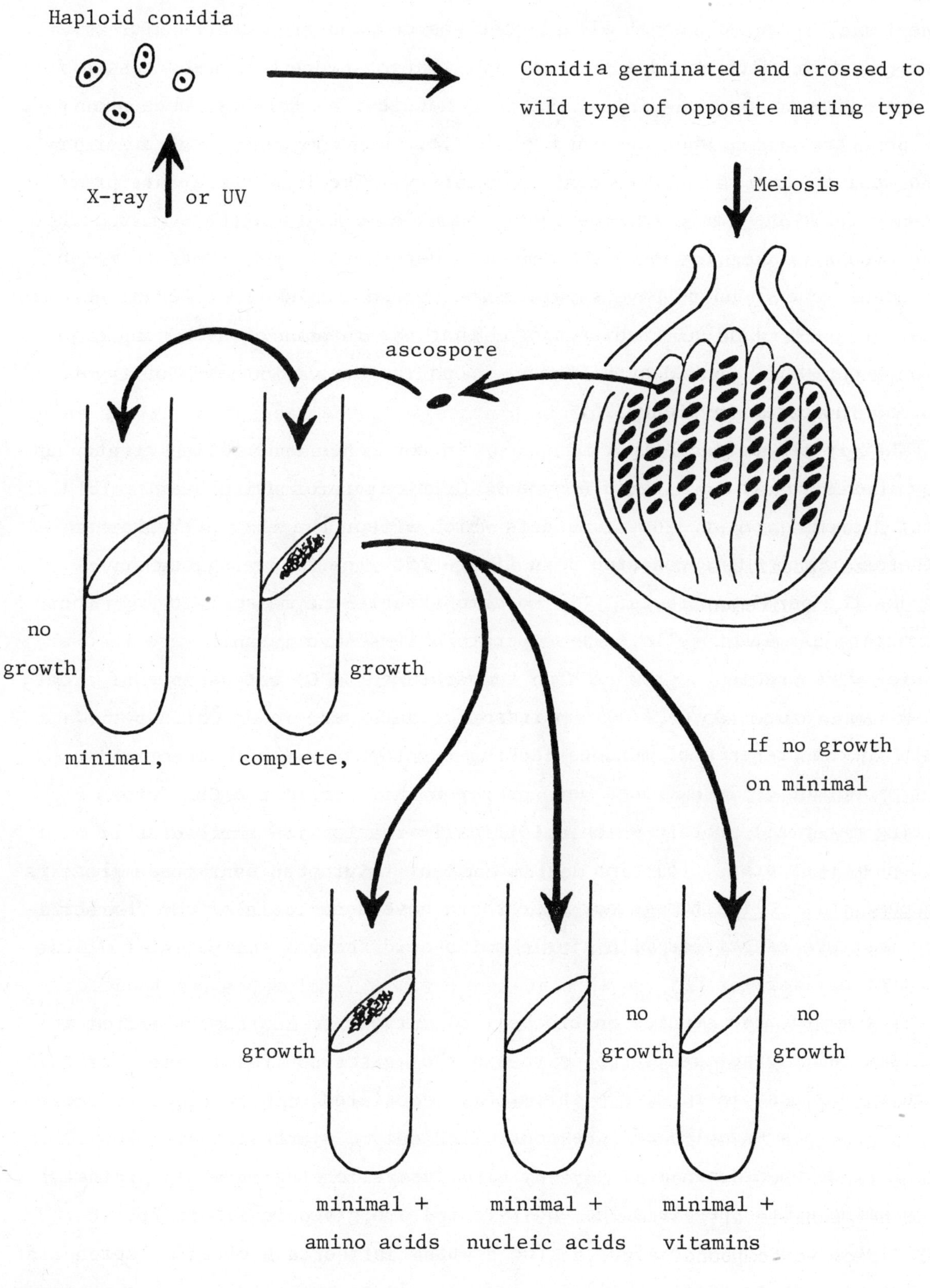

Conclusion: amino acid mutant recovered.

are genetically and biochemically distinctive. Socalled tryptophaneless mutants (try^-), incapable of growth in the absence of the amino acid tryptophan, will serve to illustrate this situation. The conclusion that independent try^- mutants are not all alleles was deduced from the results of experiments in which conidia from two try^- mutants were mixed and tested for their ability to grow on minimal medium. Results, as follows, were obtained. (1) Some try^- pairs when combined grow, i.e., are complementary; other try^- pairs fail to grow, i.e., are noncomplementary. The former are interpreted as non-allelic try^- mutants, the latter as allelic try^- mutants. Altogether three groups of complementary try^- mutants were found. (2) Mutants which complement were found to have separate loci often linked to different chromosomes. (3) Biochemical studies showed that the mutants of each complementation group were blocked at different steps in the biosynthesis of tryptophan, as indicated in figure 25.

What is the evidence that mutations in the genes controlling tryptophan biosynthesis result in altered enzymes? Critical information bearing on this question stems from studies of mutants which affect the enzyme tryptophan synthetase which catalyzes step 3 in figure 25. The following summary includes the pertinent facts: (1) no demonstrable enzyme activity was found in extracts of mutant cells, (2) temperature sensitive mutants were recovered which were unable to grow at high temperature (35°C) but capable of growth at low temperature (25°C), (3) suppressor mutants were found which permit growth in the presence of mutants lacking tryptophan synthetase activity, (4) from mutant cells extracts were prepared which contain a CRM (cross reacting material) protein which specifically reacts with antibodies to tryptophan synthetase, (5) finally, the mutant tryptophan synthetase proteins of _Escherichia coli_ although not Neurospora have been isolated and characterized; most are associated with single amino acid changes equivalent to those described earlier for Hb.

In summary, the studies on biochemical mutants of Neurospora and microorganisms fully vindicate Garrod's thesis that genes control enzyme synthesis and these enzymes, in turn, are intimately associated with intermediary metabolism. Numerous examples of gene-controlled enzyme synthesis have been thoroughly documented in man. By way of illustration the metabolic pathways of the amino acids phenylalanine and tyrosine are given in figure 26. Hereditary blocks are known for four of the steps. In the case of phenylketonuria the block occurs in reaction (1) and individuals so affected lack the enzyme phenylalanine oxidase, excrete excessive amounts of phenyllactic and

NOTES

Figure 25. Biosynthesis of tryptophan in Neurospora

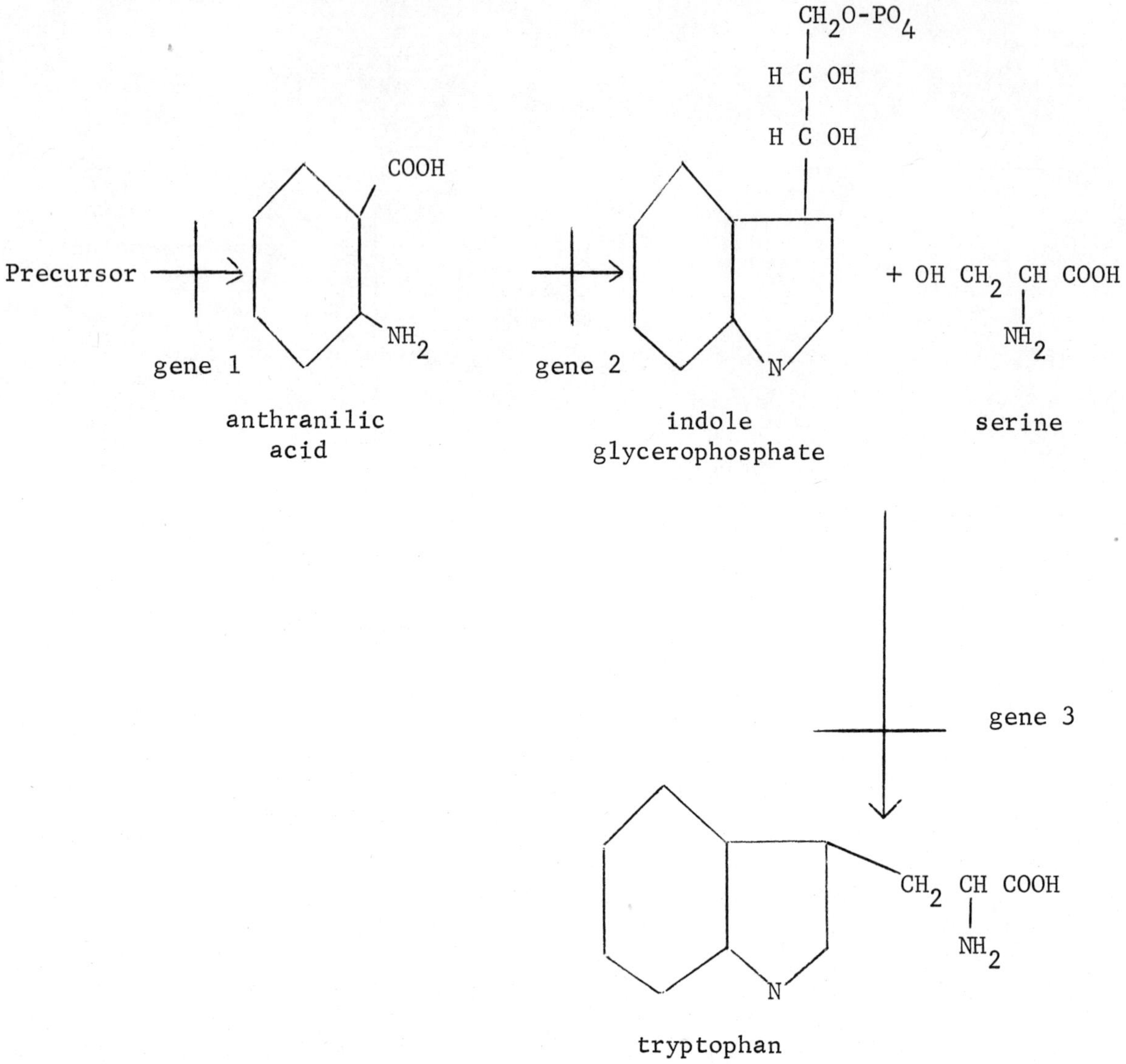

Figure 26. The metabolism of tyrosine and phenylalanine

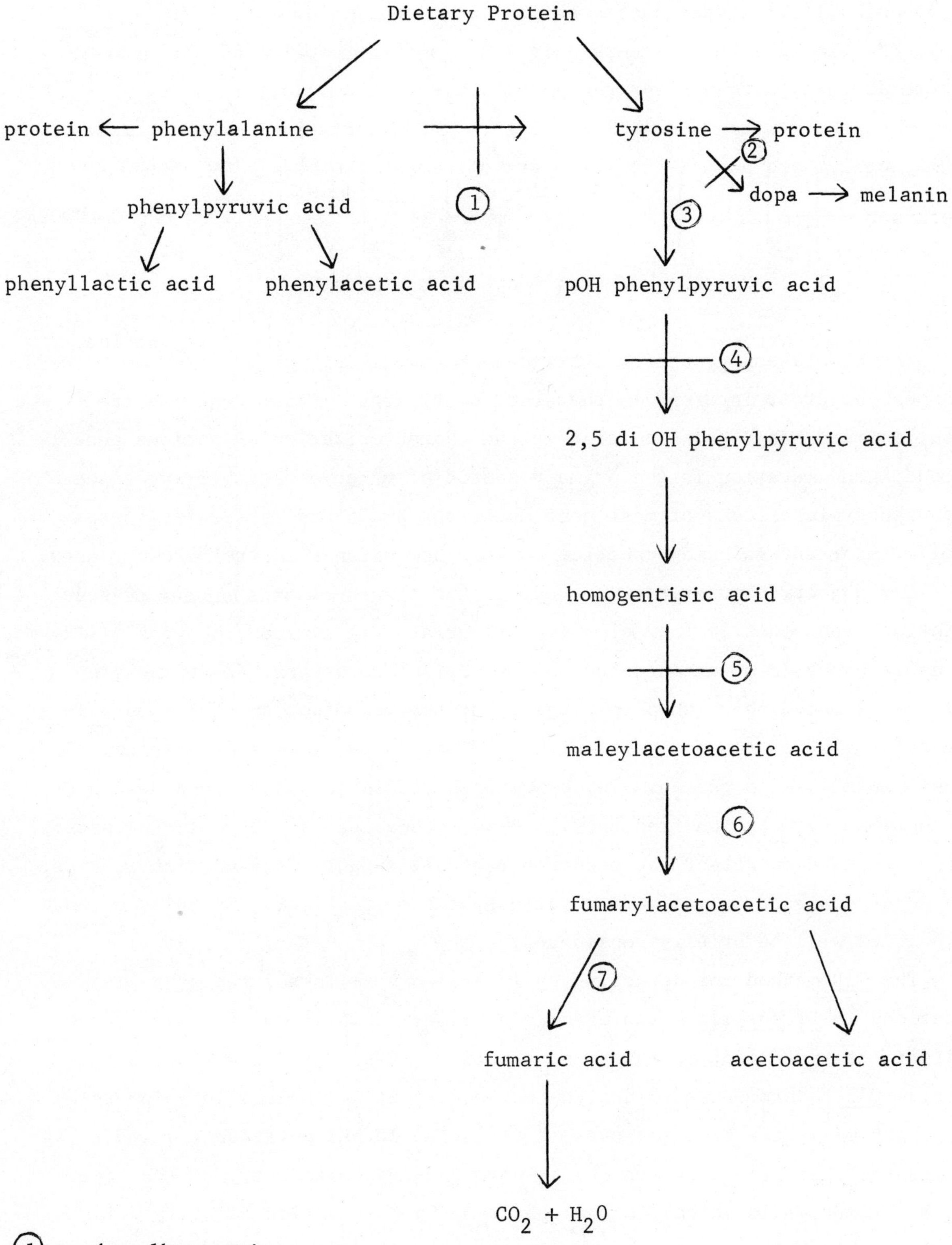

(1) = phenylketonuria

(2) = albinism

(4) = tyrosinosis

(5) = alkaptonuria

phenylpyruric acid, and accumulate the amino acid phenylalanine. If uncontrolled, mental deficiency results. In the case of alkaptonuria where the block occurs at step (5), it has been possible to perform enzyme assays on biopsied liver tissue and establish, as exemplified by the data in table 29, that all but one enzyme is functioning properly and the missing enzyme activity accounts for the accumulation of homogentisic acid.

The aforementioned examples are fully representative; inborne errors of metabolism are now an integral part of human genetics. (See table 29b.)

22. The Frequency of Gene Mutation

The spontaneous occurrence of gene mutations is, for at least two reasons, of great interest in the study of biology. First, knowing the mutation rate has intrinsic value in the characterization of various genes. Second, gene mutation is the primary source of genetic variability; since the phenotypic effects of most gene mutations are undesirable, it is especially important in human genetics to know how often such traits occur anew.

Any objective quantitative study of the frequency of spontaneous gene mutations must fulfill the following criteria: (1) exclude the possibility of differences in mutation detection by various observers, (2) establish systems of breeding which permit the quantitative detection of new mutations, and (3) allow for the analysis, with ease, of large numbers of gametes. In experimental plants and animals, a number of techniques have been designed for estimating mutation frequency. These techniques fall into two classes: those designed to detect the occurrence of all mutants of a particular type, and those designed to estimate mutations at specific loci. Examples of both techniques will be briefly considered.

The *ClB* method for detecting newly arisen sex-linked, recessive lethal mutations in *Drosophila* is an example of a method which fulfills all three criteria for quantitative mutation detection. Basic to the method is the special *ClB* X chromosome so designated because it includes a long inversion (C) which minimizes crossing over, a recessive lethal mutation (l) and a dominant marker gene, Bar eye (B). The method exploits the fact that the *ClB* X chromosome is lethal to Drosophila males. Detection of newly arisen sex-linked lethal mutations is wholly objective and depends simply upon distinguishing between individual F_2 cultures containing and lacking male progeny. The latter is the consequence of a new sex-linked, recessive

Table 29. Comparative enzyme activity in livers of normal and alkaptonuric individuals (modified after La Due et al., 1958)

		Enzyme activity (arbitrary units)	
		Normal	Alkaptonuria
(3)*	tyrosine transaminase	3.6	3.2
(4)	pOH phenylpyruric acid oxidase	6.7	4.6
(5)	homogentisic acid oxidase	26.8	<0.0048
(6)	maleyl acetoacetic isomerase	960	780
(7)	fumarylacetoacetic hydrolase	29	22

*metabolic step in figure 26

Table 29b. A sample of human hereditary enzyme deficiencies

Disease	Enzyme	Phenotypic Effect
phenylketonuria	phenylalanine hydroxylase	mental retardation
alkaptonuria	homogentisitic acid oxidase	arthritis
albinism	tyrosinase	pigmentation, vision
galactosemia	galactose-1-phosphate uridyl transferase	mental retardation
favism	glucose-6-phosphate dehydrogenase	hemolytic anemia
Tay-Sachs	hexosaminidase A	progressive neurological deterioration
Lesch-Nyhan	hypoxanthine-guanine phosphoribosyl transferase	mental retardation, self-mutilation
maple syrup urine	keto acid decarboxylase	mental and physical retardation
congenital lactose intolerance	lactase	chronic diarrhea
tyrosinemia	pOH phenylpyruvic oxidase	liver damage

lethal mutation which occurred in a parental male. The procedure is outlined in figure 27. As noted the method is entirely objective and since large numbers of individual F_1 females can be tested by crosses in small culture vials, literally thousands of X chromosomes can be easily assayed.

In maize and the mouse, specific locus techniques have been devised which allow for the testing of mutations at several loci simultaneously.

Estimating mutation rates in man is fraught with difficulties. The inability to control matings is especially disadvantageous. Since procedures designed to estimate mutants at specific loci are not possible, the detection of new mutations is primarily phenotypic. This raises a number of questions which have to be resolved if a mutation estimate is to be reliable. (1) Is the phenotype being scored characteristic of mutations at a single locus or do mimics occur because of mutations at other loci? In principle this depends to a large degree on the precision with which the phenotype can be defined. Thus, for example, if impaired blood coagulation is the phenotype then new mutations of hemophilia A and hemophilia B would be scored as one entity. By resorting to a detailed study of the impaired blood coagulation mechanism, mutations of each gene can be distinguished. (2) Is the penetrance complete? Without this information, it is not always possible to objectively distinguish between a new mutation and an existing mutant segregating from one or the other parent where the mutant phenotype was masked by incomplete penetrance. (3) Is the mutation germinal or is the newly arisen phenotype the result of either a somatic mutation or a phenocopy? The critical test of a germinal mutation is its transmission to the progeny of the bearer. There is evidence, to be considered shortly, that somatic mutations and/or phenocopies do occur and thus mutation rates based solely on phenotype are often overestimates. (4) Is the phenotype the consequence of a bona fide gene mutation or the product of a chromosomal aberration? For the time being cytogenetic methods in man are insufficient to distinguish between the two possibilities. Based on experience with experimental animals and plants, there is little doubt that often supposed gene mutations are in reality chromosome aberrations, e.g., deficiencies, duplications, etc. (5) Is the mutation rate representative? Obviously mutation rates based on phenotypic methods are those most convenient to estimate. Where the frequency is rare (e.g., $1/10^7$ births) or the phenotypic changes slight, analysis is difficult and will rarely be undertaken.

Figure 27. The ClB method for detecting sex-linked recessive lethal mutations in Drosophila

P ♂ +/Y X ClB/+

F_1 progeny = ♀♀ ClB/+; +/+; ♂♂ +/Y ClB/Y dies

F_1 ClB ♀♀ crossed individually to F_1 ♂♂ thereby testing single X chromosomes derived from a P ♂

If an F_1 ClB ♀ carries a new lethal mutation 1^x, which arose in the P ♂

then, ClB/1^x X +/Y → F_2 ♀♀ ClB/+, +/1^x; ♂♂ ClB/Y die; 1^x/Y die

sex ratio = 2 ♀:0 ♂

If an F_1 ClB ♀ carries an unchanged paternal X chromosome

then, ClB/+ X +/Y → F_2 ♀♀ ClB/+, +/+; ♂♂ +/Y, ClB/Y dies

sex ratio = 2 ♀:1 ♂

Direct determination of mutation frequency: In principle, this method applicable primarily to dominant mutations is simple. The progeny of parents not showing a particular trait are censused for the dominant trait. For the trait involved it is assumed that (1) only one locus is involved, (2) penetrance is complete, (3) the trait is never mimicked by recessive genes, (4) all new occurrences of the trait are the result of germinal mutations. The mutation rate μ equals $n/2N$ where n is the number of new mutants and N is the number of births censused. In table 30 several independent mutation rate estimates by the direct method are listed. One comment is in order: note that one retinoblastoma mutation rate estimate is corrected for the non-transmissible types. This suggests that other retinoblastoma mutant rates are overestimates and for that matter all the other rates in table 30 could be too high.

Indirect determination of mutation frequency: The rationale of this method is straightforward. Assume that an equilibrium exists between the rate of occurrence of new mutations and the rate of elimination of already existing mutants because of their reduced fitness (selective disadvantage). Reduced fitness or selective disadvantage means that a mutant allele is less likely to be transmitted to the next generation than is its normal allele. Equilibrium is maintained by the mutation rate which compensates for the loss of alleles because of reduced fitness. For dominant genes the mutation rate can be estimated from the following:

Let N = the number of people in a prescribed population
μ = the mutation rate
then $2N\mu$ = the number of new dominant genes.

Let f = the reduction in fitness due to the dominant gene
$s = (1-f)$ = the proportional reduction in fitness due to the dominant gene
x = the number of individuals with the dominant gene phenotype

at equilibrium $2N\mu = Nxs$
or $\mu = xs/2.$

An example of the application of this method is given in table 31 where the mutation rate for the dominant gene causing chondrodystrophy is estimated.

Table 30. Estimates dominant, autosomal gene mutation rates ("direct method") (modified after Neel, 1962)

Trait	Rate/10^6 loci/generation	
Epiloia	8	
Chondrodystrophy	45	(Denmark)
	70	(Sweden)
	143	(N. Ireland)
	120	(Japan)
Aniridia	5	(Denmark)
	5	(U.S.A.)
Pelger anomaly	27	(Germany)
	17	(Japan)
Retinoblastoma	12	(England)
	23	(U.S.A.)
	18	(U.S.A.)
	21	(Japan)
	21	(Switzerland)
	4*	(Germany)

*Corrected for nontransmissible types

Note that this estimate compares favorably with the estimates by the direct method for Denmark and Sweden in table 30. These results suggest further that the chondrodystrophy rates for North Ireland and Japan in table 30 are very likely too high.

The thinking underlying the indirect method of mutation rate estimates for dominant genes applies equally well for autosomal, recessive genes. At equilibrium the following relationships will hold:

$$2N\mu = 2Nxs$$
$$\mu = xs$$

Note that it is assumed that the fitness of heterozygotes is equal to that of homozygotes for the normal allele. The plausibility of this assumption can be questioned at least for some cases. For example, there is good reason to believe that in regions where malaria is endemic sickle cell trait individuals are fitter than normal individuals. Under this condition, the mutation rate of the sickle cell gene would be overestimated. Nonetheless, estimates of mutation rates have been made for several traits including albinism, total colorblindness, amaurotic idiocy, amyotonia congenita, microcephaly and phenylketonuria. All give rates of 1-5 X 10^{-5}.

The indirect method has also been used to estimate the frequency of sex-linked mutations in males. Here the equilibrium relationship becomes

$$\mu = sx/3.$$

It is assumed that selection against females heterozygous for a sex-linked recessive is neglible compared to selection against the hemizygous male. By the indirect method two independent estimates of the mutation rate of hemophilia gave 3.2 X 10^{-5} in Denmark and 2.7 X 10^{-5} in Switzerland. Both are not unreasonable estimates.

Mutagenic agents: Since the discovery in 1927 that the irradiation with X-rays of Drosophila and several higher plants significantly increases the frequency of all kinds of gene mutations and chromosomal aberrations, a wide variety of mutagens, agents capable of increasing the mutation rate, have been described. These include other kinds of high energy radiation, e.g., neutrons and gamma rays, ultraviolet light and a variety of chemical mutagens many of which are extraordinarily potent. Fortunately, the bulk of the

Table 31. Estimate of mutation to chondrodystrophic dwarfism by the indirect method

$$\mu = \frac{sx}{2}$$

(μ = mutation rate, s = proportional reduction in fitness due to dwarfism gene, x = frequency of dwarf phenotype)

Estimated frequency of dwarfism = 10/94,075 births

Estimated fitness of dwarfs:

108 dwarfs produced 27 children (10 dwarfs, 17 normal) or on the average 13.5 dwarfism genes were transmitted to next generation, therefore, proportion of dwarfism genes transmitted = 13.5/108 = 0.125.

457 normal sibs of dwarfs produced 582 children or of available non-dwarf alleles (457 X 2 = 914), 582 transmitted to next generation, therefore, proportion of non-dwarf genes transmitted = 582/914 = 0.6368.

$$\text{Relative fitness of dwarfs} = \frac{0.125}{0.6368} = 0.1963.$$

Thus, s = reduced fitness because of dwarfism gene = 1-0.1963 = 0.804.

$$\text{Thus, } \mu = \frac{0.804 \text{ X } 10/94{,}075}{2} = 4.27 \text{ X } 10^{-5}$$

mutagen research has been done on experimental organisms, e.g., *Drosophila*, mice, *Zea mays* and microorganisms and, in fact, there is little direct information on mutagen effects in man. Nonetheless, there is every reason to believe that the human genome is as susceptible to mutagenic agents as are experimental organisms. Studies with human cells in tissue culture demonstrates that aberrations of human chromosomes are readily inducible with both high energy irradiation and chemical agents.

23. Polygenic Inheritance or The Genetics of Continuous Variation

The contrasting traits considered thusfar, those governed by the rules of Mendelian genetics, are said to manifest discontinuous variation. This means that for a given population, each individual can be assigned to one of a limited number of discrete phenotypic classes. It is obvious that not all contrasting traits can be easily fitted into a finite number of phenotypic classes. Included here are metrical traits such as stature, weight, intelligence which are said to manifest continuous variation.

Size inheritance in the rabbit exemplifies the complications encountered when studying the inheritance of a trait showing continuous variation. Among the breeds developed through selection by rabbit fanciers are the small Polish (mean weight 5-7 kilo) and the Flemish giant (mean weight 25-30 kilos). Crosses between the two breeds produce hybrid individuals of intermediate size (mean weight 15-18 kilo). However, the F_2 individuals produced by crosses between the hybrid individuals cannot be conveniently fitted into the three size classes of the parents and their hybrid. Rather their sizes are distributed more or less continuously between the sizes of the parental breeds according to a normal or Gaussian frequency distribution. How may these results be explained within the framework of Mendelian genetics? At a minimum, three assumptions are required. (1) A comparatively large number of gene loci (polygenes) are involved in the determination of the phenotype, size. (2) The contributions of each locus are equal, i.e., their contributions to the phenotype are additive. (3) The effects of the environment on the phenotype are equally distributed among all the possible genotypes.

That the environment is a source of a substantial portion of phenotypic variability follows from the classical experiments of Johannsen on the sources of seed size variability in the garden bean. In the Danish environment where Johannsen conducted his experiments, the garden bean is effectively 100%

NOTES

self-pollinated and the small sample of beans with which the experiments were initiated was expected to be homozygous at virtually all loci. A substantial amount of variation was observed within the unselected seed sample for individual seeds ranged in weight from ca. 20 to more than 70 centigrams per seed. When the seed weights of the offspring of these seeds were determined, the results obtained were those given in table 32. Note that small seeds tend to produce small seeds and large seeds tend to produce large seeds. In addition, the mean weight of the progeny is related to that of the parent although the mean differences among the offspring are less than those among the parents. This means that seed size, a trait of continuous variation, has a genetic component. There is, in addition, a large environmental component which accounts for much of the seed size variation. Johannsen extended his studies by first establishing 19 lines, each originating from a different individual, and then analyzing the variation in seed size in each line. Representative lines and the seed weights of parents and progeny are given in table 33. Note that each line has its characteristic seed weight and that within a line irrespective of the weight of the parent seeds the seed weight of their offspring is the same.

In subsequent experiments Johannsen selected heavy and light seeds within each line to test whether the within line variation was heritable or not. After six generations of selection no significant differences were achieved which means that the within line differences are environmental effects. In summary, Johannsen's experiments established the following points: (1) a trait can simultaneously show genetic and environmental variation, (2) by appropriate breeding experiments, the two sources of variation can be distinguished, (3) the graded series from light to heavy seeds with no clear-cut discontinuities shows that there are many genes with slight additive individual effects involved, (4) the environmental effects can in some cases reinforce, in some cases oppose the genetic differences.

It is clear from the foregoing results that variation in a metrical trait is attributable to two major components: the genotype and the environment. The first objective in the study of this variation is to assess the relative magnitude of each component. In other words, a rational answer is sought to the longstanding nature-nurture question. For a polygenic trait where the effects of the genes are equal and additive and where the environmental influences are trivial, the degree of resemblance between relatives should be correlated with the number of genes they share in common, or the proportion of their genes derived from a common ancestor. Table 33 outlines this

Table 32. Seed weights of parents and progeny in the garden bean (after Johannsen)

Centigram Weight Parent	Weight of Progeny									
	10	20	30	40	50	60	70	80	90	Mean
20		1	15	90	63	11				43.8
30		15	95	322	310	91	2			44.5
40	5	17	175	776	956	282	24	3		46.2
50		4	57	305	521	196	51	4		48.9
60		1	23	130	230	168	46	11		51.2
70			5	53	175	180	64	15	2	56.0
Total	5	38	370	1676	2252	928	187	33	2	47.9

Table 33. Seed weight of parents and progeny within several lines of the garden bean (after Johannsen)

Centigram Weight	Mean weight of progeny in line				
Parent	1	2	13	18	19
20				41.0	
30			47.5	40.7	35.8
40		57.2	45.0	40.8	34.8
50		54.9	45.1		
60	63.1	56.5	45.8		
70	64.9	55.5			
Line mean	64.2	55.8	45.4	40.8	35.1

relationship between relatives. The metrical trait of fingerprint ridge counts illustrates a polygenic trait where the genotype is preeminent. The number of ridges is determined by the fourth fetal month and, therefore, the environmental effects should be minimal. In table 35 both the observed and expected ridge count correlations between relatives are given. These results establish the intimate association between the proportion of genes held in common and the phenotypes of related individuals.

Biometrical methods designed to estimate the contributions of the genotype and the environment of the phenotype of metrical traits start with the premise that G + E = P (where G, E and P are the variations in genotype, environment and phenotype, respectively). By definition, the heritability or fraction of the phenotypic variation due to genetic variation equals G/P and is commonly designated h^2. That fraction of the phenotypic variation due to variations in the environment equals E/P or e^2. Thus,

$$h^2 + e^2 = 1.$$

It follows that h^2 can vary from 1 where no environmentally caused variation occurs to 0 when all the variation is environmental. In principle, h^2 is a measure of the correlation between genotype and phenotype in a proscribed population. Biometrical methods of estimating h^2 are based on the idea of genetic correlation between related individuals, e.g., parent and offspring or between siblings or between other individuals who on the average have the same genes in common.

One approach at estimating the genetic and environmental components of phenotypic variation uses monozygotic (identical) and dizygotic (fraternal) twins. In principle, the analysis is simple. Since monozygotic twins (MZ) have identical genotypes, phenotypic differences between MZ pairs must be due almost entirely to environmental differences. In the case of dizygotic twins (DZ), pairs may differ both because of different genetic endowments and environmental differences. Comparisons can be made both for simple Mendelian and metrical traits. For a given trait where both partners of a twin pair show the trait they are called concordant, but when only one of the partners shows the trait they are discordant. A sampling of twin concordance data is given in table 36 for which a few comments are in order. Caution must be exercised in interpreting the concordance data. In the case of infectious diseases, e.g., tuberculosis and rheumatic fever an interpretation that the difference between MZ and DZ twins is due to their heredity

Table 34. Genetic relationships between relatives

Relationship to a specific individual	Proportion of genes in common
Identical twins	1
Parent-child; sibs fraternal twins	1/2
Grandparent-grandchild; Uncle/aunt-nephew/niece; half-sibs; double first cousins	1/4
First cousins	1/8
Second cousins	1/32

Table 35. The correlation among relatives of fingerprint ridge counts (after Holt, 1955)

	Correlation	
Relationship	Observed	Expected
mother-child	0.48 ± 0.04	0.5
father-child	0.49 ± 0.04	0.5
sib-sib	0.50 ± 0.04	0.5
identical twins	0.95 ± 0.01	1.0
fraternal twins	0.49 ± 0.08	0.5

must be accepted with reservation. The high concordance of MZ twins could very possible be the result of more intimate contact between MZ partners. For the other data, excepting coronary occlusion, there is a strong suggestion that an important hereditary component exists in the etiology of each trait. However, the twin data permit no insights into the kind and number of genetic factors involved and this remains a serious limitation in the usefulness of the twin method.

Table 36. A sampling of traits for concordance among twins (from Harvald and Hauge, 1965)

Trait	Concordance Frequency MZ	DZ_1	DZ_2
Down's Syndrome	6/6	0/8	0/15
Mental deficiency	12/18	0/20	0/29
Epilepsy	10/27	6/43	4/57
Schizophrenia	4/9	4/33	2/29
Coronary occlusion	20/102	24/155	10/133
Arterial hypertension	20/80	10/106	4/106
Tuberculosis	50/135	46/267	36/246
Rheumatic fever	30/148	12/226	14/202

DZ_1 = like sexed
DZ_2 = unlike sexed

NOTES

PROBLEM SET 1 (apropos sections 3-5)

1. The chromosome complement of an organism consists of 10 chromosome pairs. Give the chromosome number for each of the following cells:

   (a) a somatic cell at metaphase in mitosis.
   (b) a primary spermatocyte at late prophase.
   (c) a primary oocyte at metaphase.
   (d) a secondary spermatocyte at metaphase.
   (e) a secondary oocyte at telophase.
   (f) a spermatid.
   (g) an ovum.

2. Given an individual whose somatic cells contain two pairs of chromosomes. What proportion of the gametes formed by this individual will be exclusively maternal in origin, exclusively paternal in origin?

   If an individual's somatic chromosome complement consists of 6 pairs of chromosomes, what proportion of the gametes will be exclusively maternal in origin, exclusively paternal in origin?

3. Figure i is a kindred of deaf-mutism associated with albinism (Tietz, Am. J. Human Genet. 15:259, 1963). Based on the information of this kindred, what is the mode of inheritance of this type of deaf-mutism? Give the genotypes of the twins and their sibs in generation VI. If each twin marries, what is the probability each will transmit the deaf-mutism trait to his progeny?

Figure i

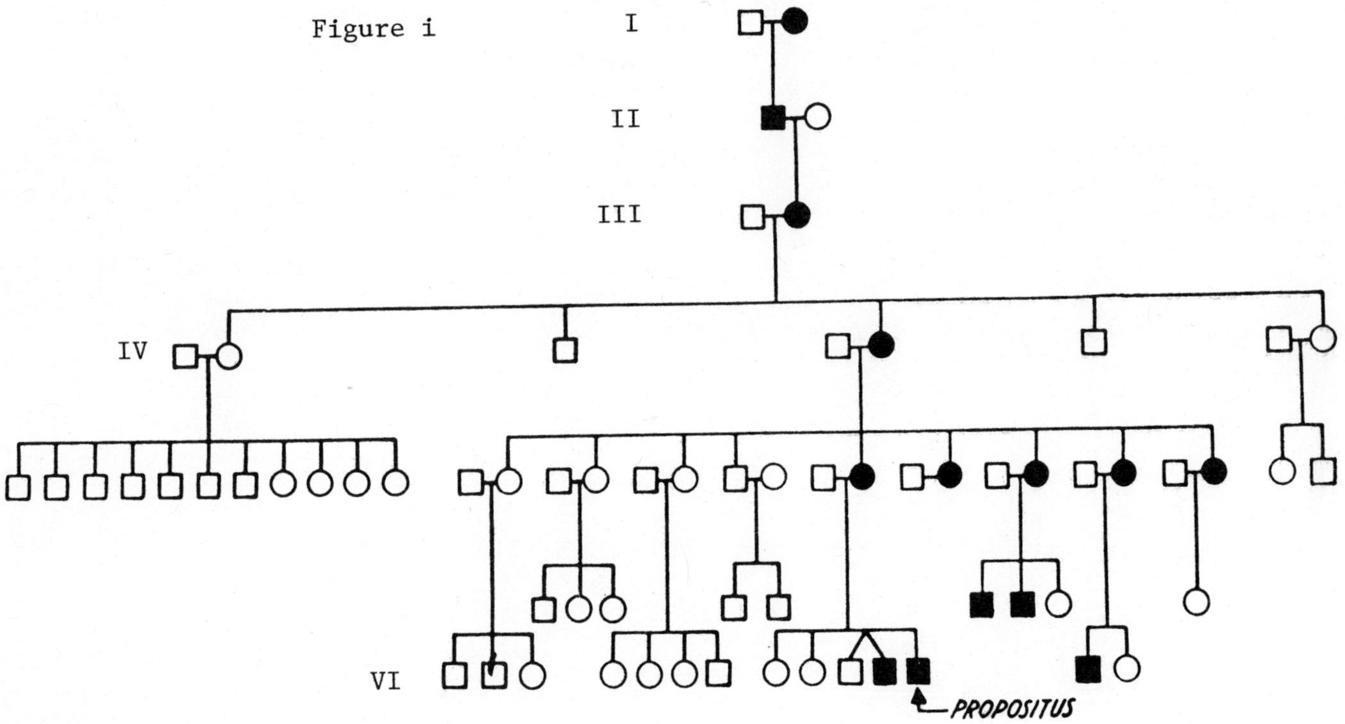

NOTES

4. In the following kindred <u>♂♀</u> = albino, ♂♀ = normal pigmentation

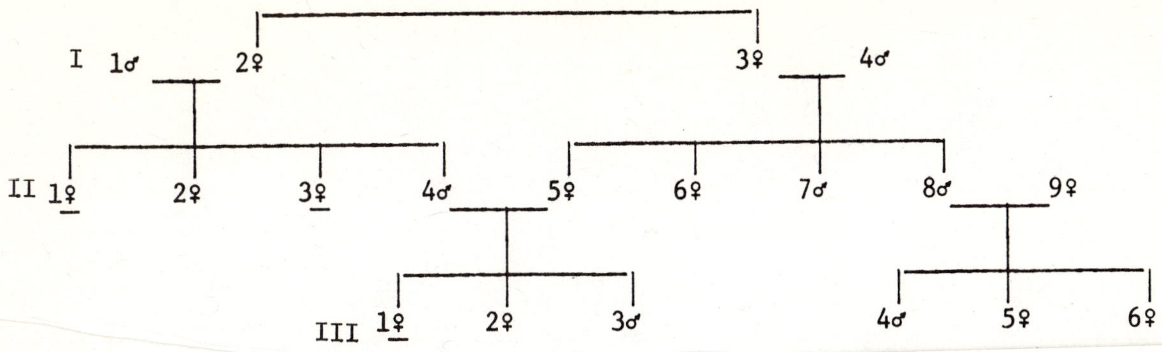

Based only on the information of this kindred, what individuals are known to be homozygous? heterozygous? Among the progeny of the following marriages, what proportion are expected to be albino: ♀II-1 X ♂II-7, ♀III-2 X ♂III-4?

5. Figure iii below represents a kindred of the socalled Stoltzfus (St) blood group (Bias, et al., Am. J. Human Genet. 21:552, 1969). What is the mode of inheritance of this blood group? What is the genotype of IV-1, IV-3, V-1? What is the probability that IV-5 is a homozygote?

Figure iii

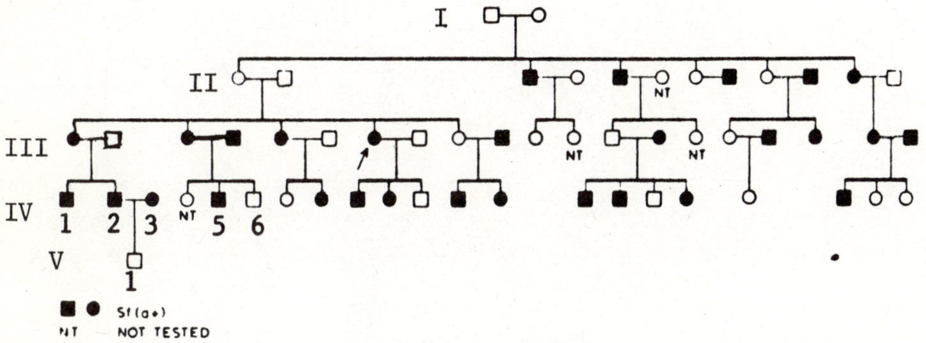

NOTES

PROBLEM SET 2 (apropos section 5)

1. Two individuals with the sickle cell trait marry. List the phenotypes with frequencies of their children. Suppose their first child is normal, what is the probability their second child will be normal? That both children will bear the trait? That one will be anemic, one normal?

2. Two normal persons produce an albino child. What is the probability their second child will also be an albino? Will have normal skin color? Give the genotypes of all persons mentioned.

3. Complete the following table computing the expectations for each of the blank spaces.

Parental phenotypes	Total children	Number of children expected M	MN	N
M X M	120			
M X MN	280			
M X N	128			
N X N	225			
N X MN	220			
MN X MN	250			

4. Refer to figure 6, the kindred of non-allelic albinism. Suppose two people whose genotypes are identical to that of ♂IV-3 marry. Give the genotypes with frequency of their expected progeny.

5. In the mouse coat color may be agouti, black or albino. Three crosses were made; assume all parents are homozygous. In cross 1, agouti was crossed to black with all F_1 agouti in phenotype. An F_1 X F_1 cross resulted in 99 agouti and 33 black mice. Cross 2 was made between black and albino mice with the F_1 all black. An F_1 X F_1 cross produced 75 black and 25 albino offspring. Cross 3 consisted in agouti X albino. All F_1 were agouti. An F_1 X F_1 cross gave an F_2 of 81 agouti, 27 black and 36 albino mice.

   (a) Give the genotypes of all agouti, black and albino mice mentioned.

   (b) If an F_1 agouti (cross 3) is crossed to a P albino (cross 3) what phenotypes with frequencies are expected among their progeny?

NOTES

PROBLEM SET 3 (apropos section 6)

1. Members of the Blackfeet indian tribe belong to either blood group A or O conditioned by the autosomal alleles I and i. Assume all parents are heterozygous. In a family with 5 children what is the probability that the first and last children will belong to blood group O, blood group A? What is the probability this family consists of 2 O and 3 A children? Of 3 O and 2 A children?

2. A brachydactylous man marries a normal woman; they have 4 children. What is the probability this sibship consists of 1 brachydactylous and 3 normal children? Of the first child normal, the remaining 3 brachydactylous? Suppose the first child of this marriage has alkaptonuria (note: both parents normal for this trait), what is the likelihood the next child will be both brachydactylic and alkaptonuric?

3. A class in human genetics consisted of 78 ♀♀ and 22 ♂♂. A census revealed the ♀♀ to have 108 sisters and 106 brothers and the ♂♂ to have 29 sisters and 19 brothers. Apply the rationale of the Weinberg proband method and determine the correct sex ratio for this population.

4. The following frequencies of albinos among sibships of size 3 derived from normal parents were observed:

albino/sibship	no. sibships	total normal	total albino
1	27	54	27
2	9	9	18
3	1	0	3
Σ	37	63	48

Using the Weinberg proband method, compute the "sibship score" for each sibship and the overall phenotypic ratio of normal:albino.

5. The incidence of recessive, autosomal Friedrich's ataxia among the offspring of two normal parents for sibships of various sizes is given in the table below. For each sibship compute the expected frequency of ataxic individuals.

Sibship size	Number families	Ataxic individuals observed	expected
2	8	9	
3	11	14	
4	9	11	
5	9	14	
6	7	15	

6. The incidence of albino progeny derived from marriages of normal X albino persons is given in the table below. For each sibship compute the expected frequency of albinos.

Sibship size	Number families	Albino progeny observed	expected
1	2	2	
2	2	3	
3	1	1	
4	3	7	
6	2	8	

NOTES

PROBLEM SET 4 (apropos section 7)

1. What is meant by the following terms: (a) Mendelian population; (b) gene frequency; (c) Hardy-Weinberg law; (d) gene pool?

2. For each of the following populations calculate the frequency of the allele a^N.

Population	Frequency of Phenotypes M	MN	N
i	25	50	25
ii	50	0	50
iii	25	10	1
iv	30	0	70
v	0	60	40

Which among these populations is in Hardy-Weinberg equilibrium? Calculate the equilibrium gene and genotypic frequencies for those not in equilibrium.

3. Blood serum of humans can be classified as esterase positive (E^+) or esterase negative (E^-). The E^+ allele is dominant to E^-. Assume the following populations are in Hardy-Weinberg equilibrium for this pair of alleles. What is the frequency of heterozygotes (E^+/E^-) in each population?

Population	No. Tested	E^+	E^-
Eskimo	160	70	90
Japanese (North)	320	240	80
Japanese (South)	250	160	90

In the case of the Japanese (South) population, (a) what is the frequency of marriages homozygous E^+ X homozygous E^-; of heterozygote X heterozygote? (b) What proportion of the progeny derived from marriages of E^+ (phenotype) X E^- are expected to be of the E^- phenotype? (c) What proportion of the progeny derived from marriages of E^+ X E^+ are expected to be E^- in phenotype?

4. The Blackfeet indians belong to blood groups A and O conditioned by the autosomal alleles I (dominant) and i (recessive). In the following assume the Blackfeet indians on a reservation are a Hardy-Weinberg population at equilibrium. The population consists of 2,000 individuals; 1,920 blood group A, 80 blood group O.

1) What is the frequency of the i gene in this population? 2) What is the predicted <u>number</u> of heterozygous individuals in this population? 3) What is the expected frequency of marriages of O X O in this population? 4) From marriages of A X O, what proportion of the children are expected to be of blood group O?

NOTES

PROBLEM SET 5 (apropos sections 8 - 12)

1. Among the individuals of European origin in the U.S. population, the MN blood group phenotypes occur in the following frequencies: M = 30%, MN = 50%, N = 20%. If all N individuals were sterilized before reproductive age and the residual population reproduced panmicticly, what are the predicted frequencies of the N gene and of the MN phenotype in the next generation?

2. Phenylketonuria (PKU), conditioned by an autosomal, recessive gene, occurs in a frequency of 1/40,000 children. Assuming Hardy-Weinberg equilibrium and assuming PKU afflicted individuals do not reproduce, what is the expected allele frequency in the next generation? What is the expected incidence of PKU in the next generation? How long (generations) would it take for the PKU frequency to decrease from 1/40,000 to 1/100,000?

3. Based upon the data of problem (2), compute the expected frequency of phenylketonura derived from first cousin marriages and compare with the incidence in the population as a whole.

4. In the house mouse, coat color may be agouti or white-spotted. From crosses between individuals of the two phenotypes, the following results were obtained:

agouti X agouti	= all agouti
agouti X white-spotted	= 1,383 white spotted, 1,452 agouti
white-spotted X white-spotted	= 726 white-spotted, 372 agouti

Assign a genotype to each phenotype and explain the results observed for each cross.

5. Cystic fibrosis occurs with a frequency of 1/1600 births and is lethal to affected individuals before maturity. Assuming random mating of all individuals in the population, what is the expected frequency of the disease in the next generation? After 5 generations?

NOTES

PROBLEM SET 6 (apropos section 13)

1. A normal woman whose father is color blind marries a normal man. (a) What is the probability their first son will be color blind? Their third son? (b) If they have four sons what is the probability (a) the first will be color blind, the last two will be normal, (b) two sons will be color blind, two sons will be normal?

2. In the accompanying kindred of microphthalmia assume all spouses are normal. What is the mode of inheritance of this trait? Give the genotype of the following individuals: IV-1, IV-2, IV-3, IV-4, IV-5.

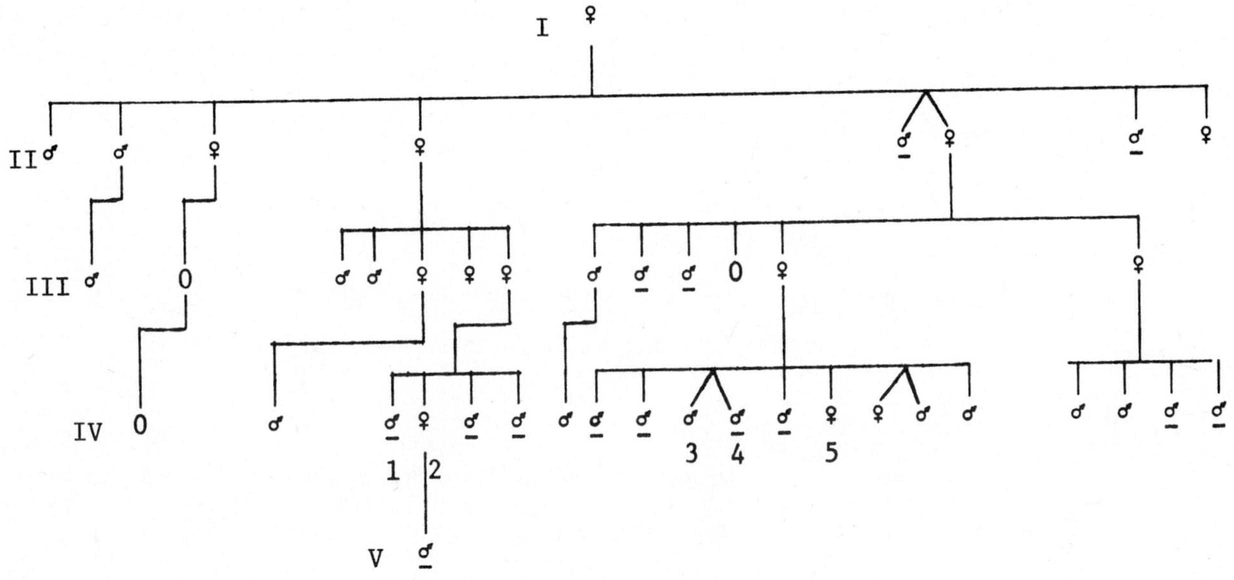

♂̲ = microphthalmia; ♂ = normal; and 0 = sex unknown

3. The inheritance of the Xg blood group is sex-linked with presence of the red blood cell antigen (Xg^+), dominant to its absence (Xg^-). The population data for the Xg blood group phenotypes among people of European origin is given below.

	Phenotype Xg^+	Phenotype Xg^-	Total
♀♀	1,488	179	1,667
♂♂	1,131	620	1,751

Based upon the data from males, (a) what is the recessive gene frequency in this population? (b) What is the expected frequency of Xg^- females? (c) What proportion of the sons of Xg^+ mothers will be Xg^- in phenotype?

4. Presence of the enzyme Glucose-6-phosphate dehydrogenase (G-6-PD) is conditioned by a sex-linked dominant gene, its absence by the recessive allele. In a Hardy-Weinberg population at equilibrium 15% of the males are G-6-PD negative. What is the expected frequency of G-6-PD negative females in this population? Of heterozygous females? Suppose all G-6-PD negative females are sterile, what is the expected frequency of G-6-PD negative males in the next generation, of G-6-PD negative females?

NOTES

PROBLEM SET 7 (apropos sections 14 and 15)

1. What is meant by the following terms: recombination, chiasma, crossing over, linkage, map location?

2. The Nail-Patella syndrome conditioned by a dominant, autosomal gene is linked to the ABO locus. Crossing over between the two loci is estimated at 14%. A woman with the syndrome belonging to the blood group A, whose father belongs to blood group O and has the syndrome, marries a normal man of blood group O. a) Give the kinds with frequency of the woman's gametes. b) What is the chance the first child of this marriage will be of the phenotype normal, A; of the phenotype normal, O? c) What is the chance the first child will be syndrome, O and the second child syndrome, A?

3. A tentative genetic map of the human X chromosome specifies hemophilia A at map location 0 and deutan color blindness at location 12. A normal woman whose mother was color blind and whose father was hemophiliac marries a normal man. What are the expected phenotypes with frequencies of the sons from this marriage?

4. Recessive protan color blindness (p vs. P), the Xg blood group (Xg^+ vs. Xg^-) and erythrocyte enzyme glucose-6-phosphate dehydrogenase (G^+ and G^-) are all sex-linked with the alleles p, Xg^- and G^- recessive. From a series of marriages of males protan Xg^+ and G^+ to females normal, Xg^-, G^- the daughters were without exception of the phenotype normal, Xg^+, G^+. These daughters produced the following sons:

$$p\ Xg^+\ G^+ = 69$$

$$P\ Xg^-\ G^- = 72$$

$$p\ Xg^-\ G^+ = 11$$

$$P\ Xg^+\ G^- = 8$$

$$p\ Xg^+\ G^- = 21$$

$$P\ Xg^-\ G^+ = 18$$

$$p\ Xg^-\ G^- = 0$$

$$P\ Xg^+\ G^+ = \underline{1}$$

$$200$$

a) What is the sequence of the three genes on the X chromosome?
b) Compute the map location for each of the three loci.

NOTES

PROBLEM SET 8

1. A panmictic population at equilibrium consists of individuals of the following phenotypes: 60% CDe, 12% cDE, 24% CDE and 4% cde. Assume allele R_1 determines CDe, R_2 determines cDE, r determines cde and no other alleles occur. What is the frequency of each of the three alleles in the population? If erythroblastosis fetalis occurs in every marriage of D/- ♂ X dd ♀♀ in which a D/- child occurs, what is the predicted frequency of the disease?

2.

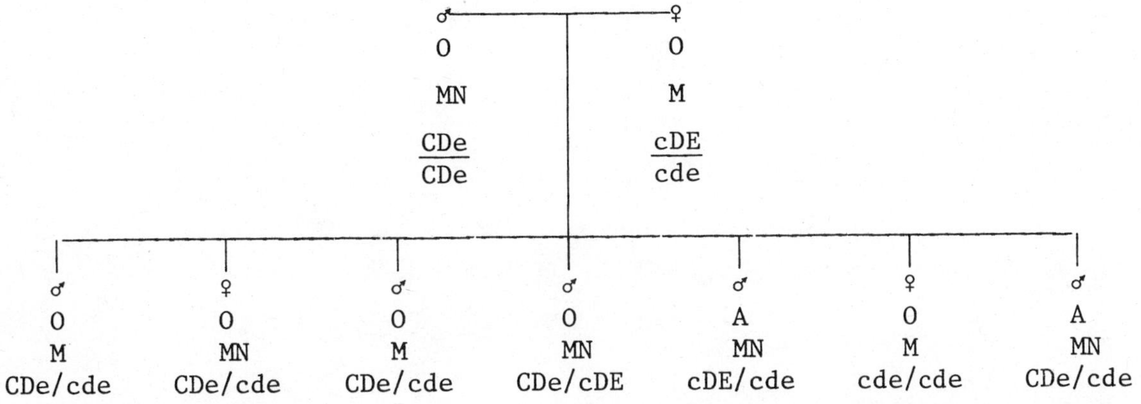

The blood groups of a purported family are given in the kindred above. Based upon this information, which children are the likely outcomes of extramarital events.

3. The Duffy blood group system specifies the antigen Fy^a and Fy^b. Absence of either antigen is designated Fy. In the table below the phenotypes of the parents and progeny of some of the possible marriages are given:

marriage	possible progeny
Fy^a X Fy^a	Fy^a, Fy
Fy^b X Fy^b	Fy^b, Fy
Fy^a X Fy^b	Fy^a, Fy^b, Fy^aFy^b, Fy
Fy X Fy	Fy only

A population censused for the Duffy phenotypes gave the following results, Fy = 9%, Fy^a = 16%, Fy^b = 55%, Fy^aFy^b = 20%. Assuming a 3 allele system and Hardy-Weinberg equilibrium, what is the frequency of the alleles in the population.

4. The ABO blood group phenotypes in a human population at Hardy-Weinberg equilibrium consists of 35% A, 24% B, 40% AB and 1% O individuals.

What is the frequency of the I^a gene in this population? What is the frequency of the i gene in this population?

NOTES

PROBLEM SET 9 (apropos section 16)

1. Two people both of the Xg^+ blood group have four children: two daughters Xg^+, one son Xg^- and a Klinefelter son Xg^-. Give the genotype of each child noting that the Xg^+ phenotype is determined by a sex-linked dominant and Xg^- by the recessive allele. What is the origin of the Klinefelter? Suppose the Klinefelter is fertile and marries a woman homozygous Xg^+. List the kinds with frequencies of the gametes formed by the Klinefelter and the phenotypes of the progeny expected from this marriage.

2. An Xg^+ woman marries an Xg^- man and their first offspring is an Xg^- Turner female. Explain the origin of the offspring. Suppose this Turner ♀ is fertile and marries an Xg^+ man. List the phenotypes with frequencies of their expected progeny.

3. In a litigation involving disputed parentage, each of two men claimed three children to be his own. The blood group phenotypes of all individuals are listed below.
husband O, MN, Rh^+; lover AB, MN, Rh^-; wife A, N, Rh^+
1st child O, MN, Rh^+; 2nd child A, N, Rh^+; 3rd child A, MN, Rh^-.
Based upon the foregoing information, which man can be EXCLUDED as the parent of the first child, the second child, the third child?

4. A 47 chromosome Down's syndrome individual marries a person with 46 chromosomes. Give the phenotypes with frequencies of their progeny.

5. A 45 chromosome normal individual marries a normal 46 chromosome individual. Their first child has Down's syndrome. What proportion of their <u>zygotes</u> can develop into a Down's offspring?

6. A 46 Down's syndrome individual marries a normal 46 chromosome individual. What proportion of their progeny are expected to be Down's?

NOTES

PROBLEM SET 10 (apropos section 18)

1. Photographs of the somatic chromosomes of five individuals (a, b, c, d, e) follow. These individuals include a normal ♂, a normal ♀, a 46 chromosome Down's individual, a 47 chromosome Down's individual and a cri-du-chat child. Determine the karyotype and the phenotype of each individual. (Cut out the chromosomes, pair them and determine the karyotype.)

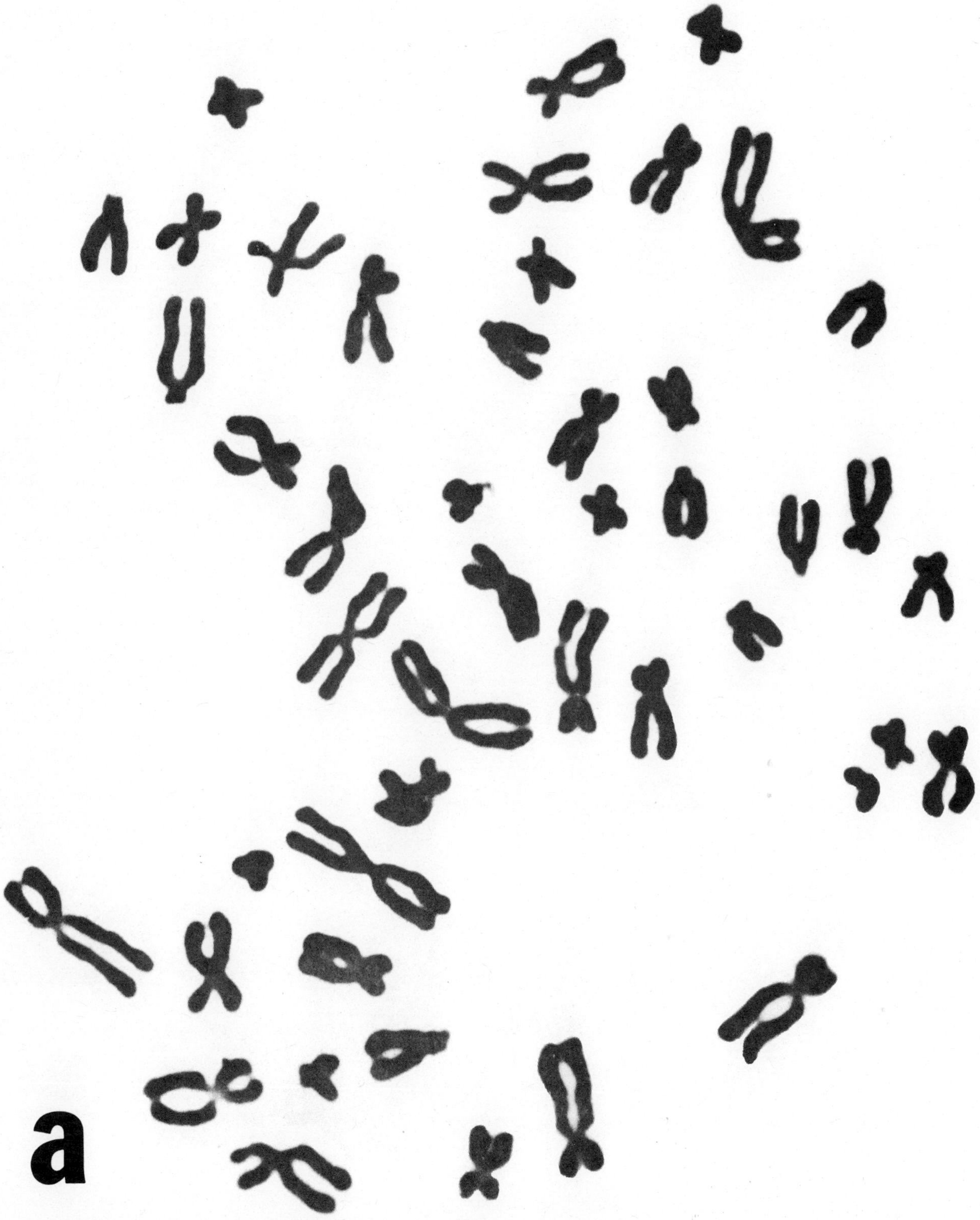

a

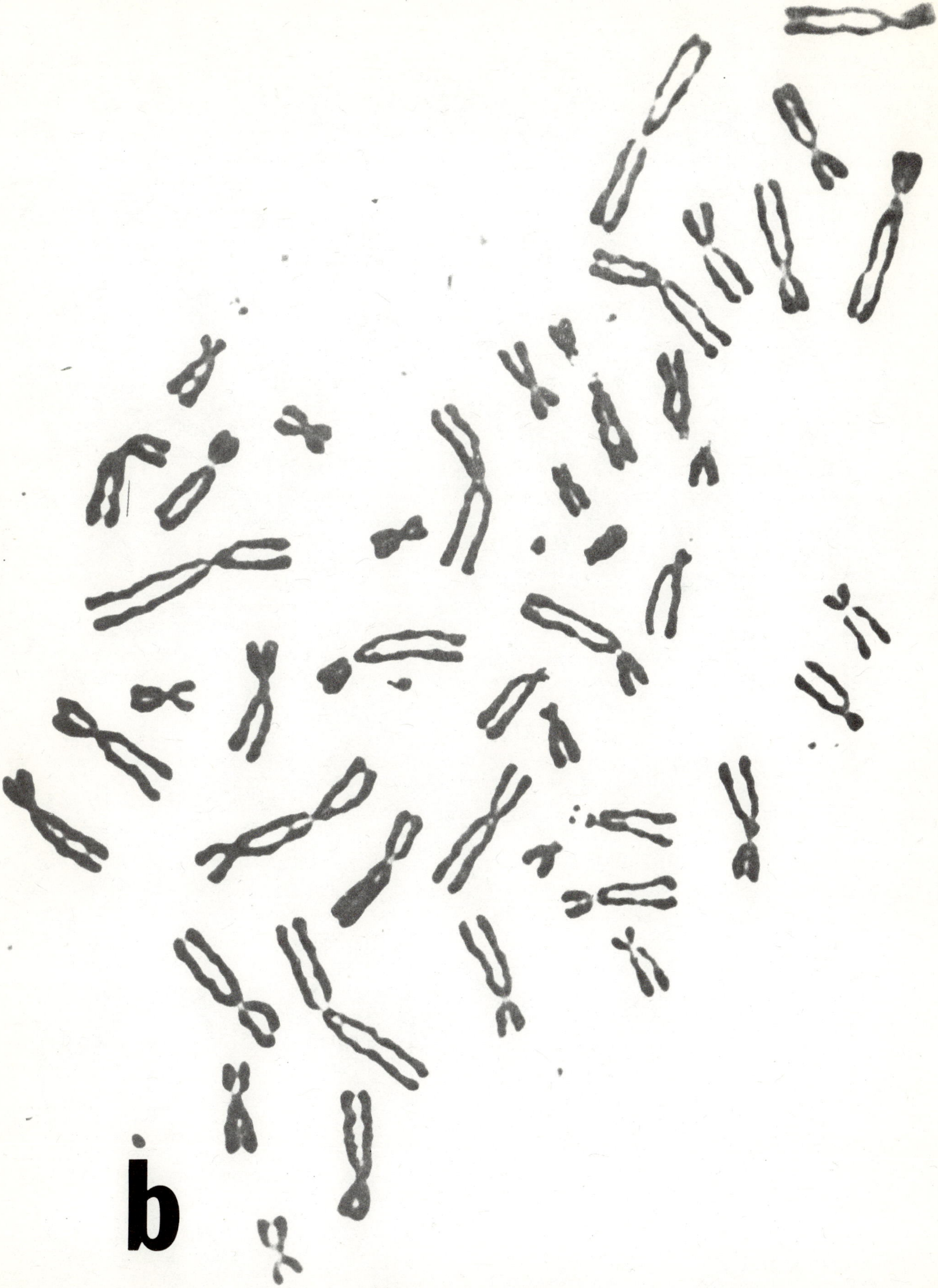

b

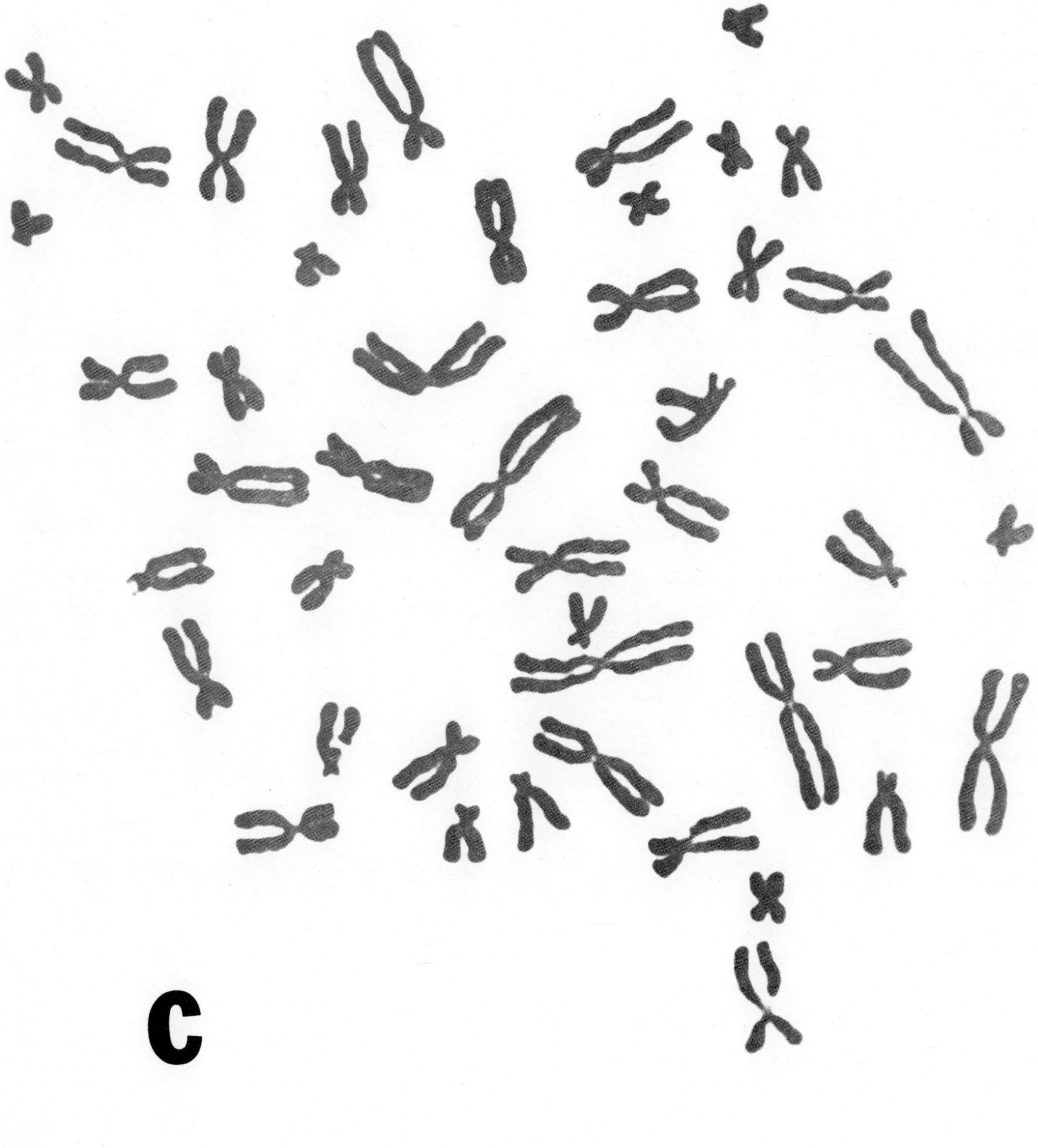

C

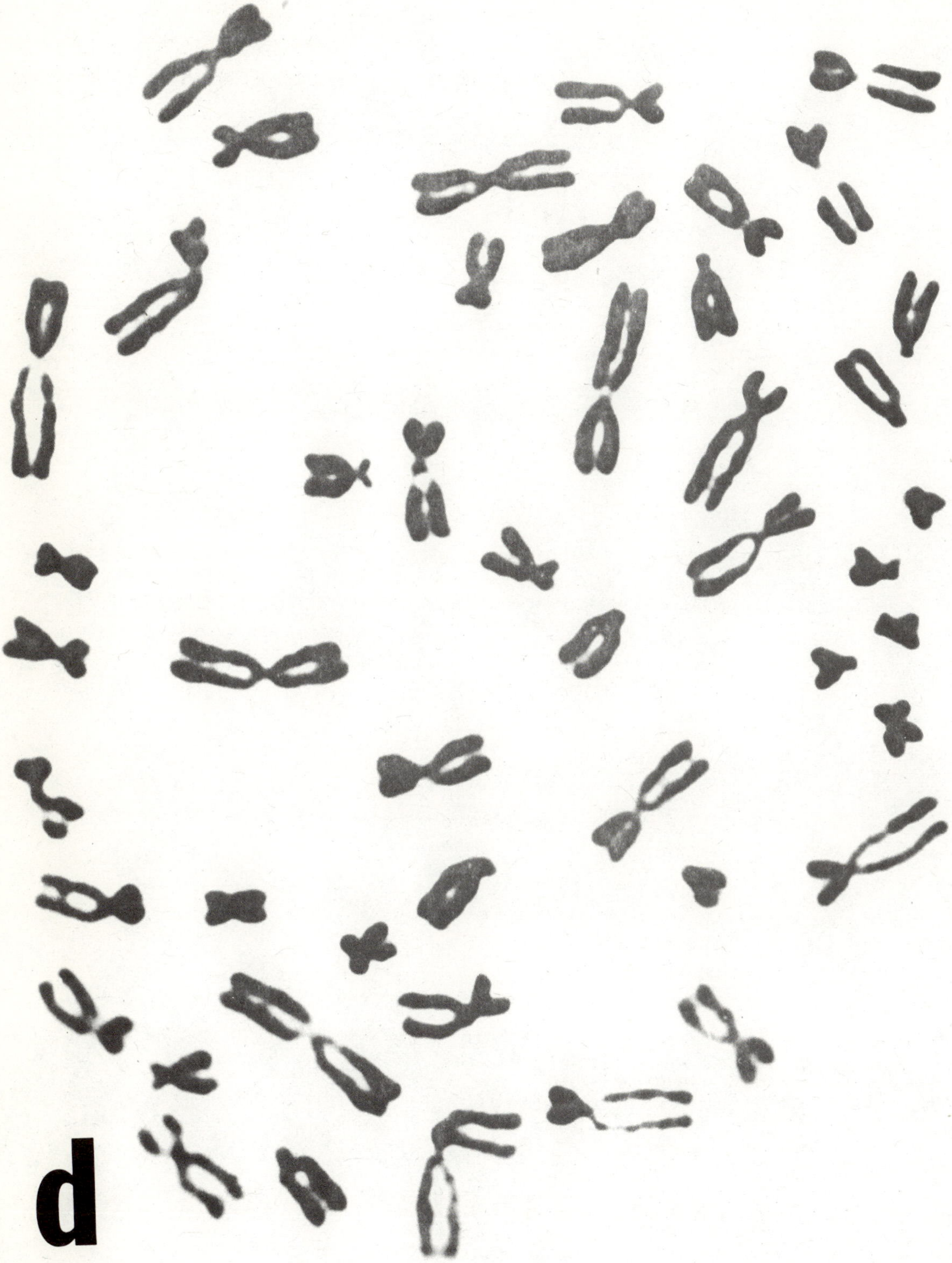

d

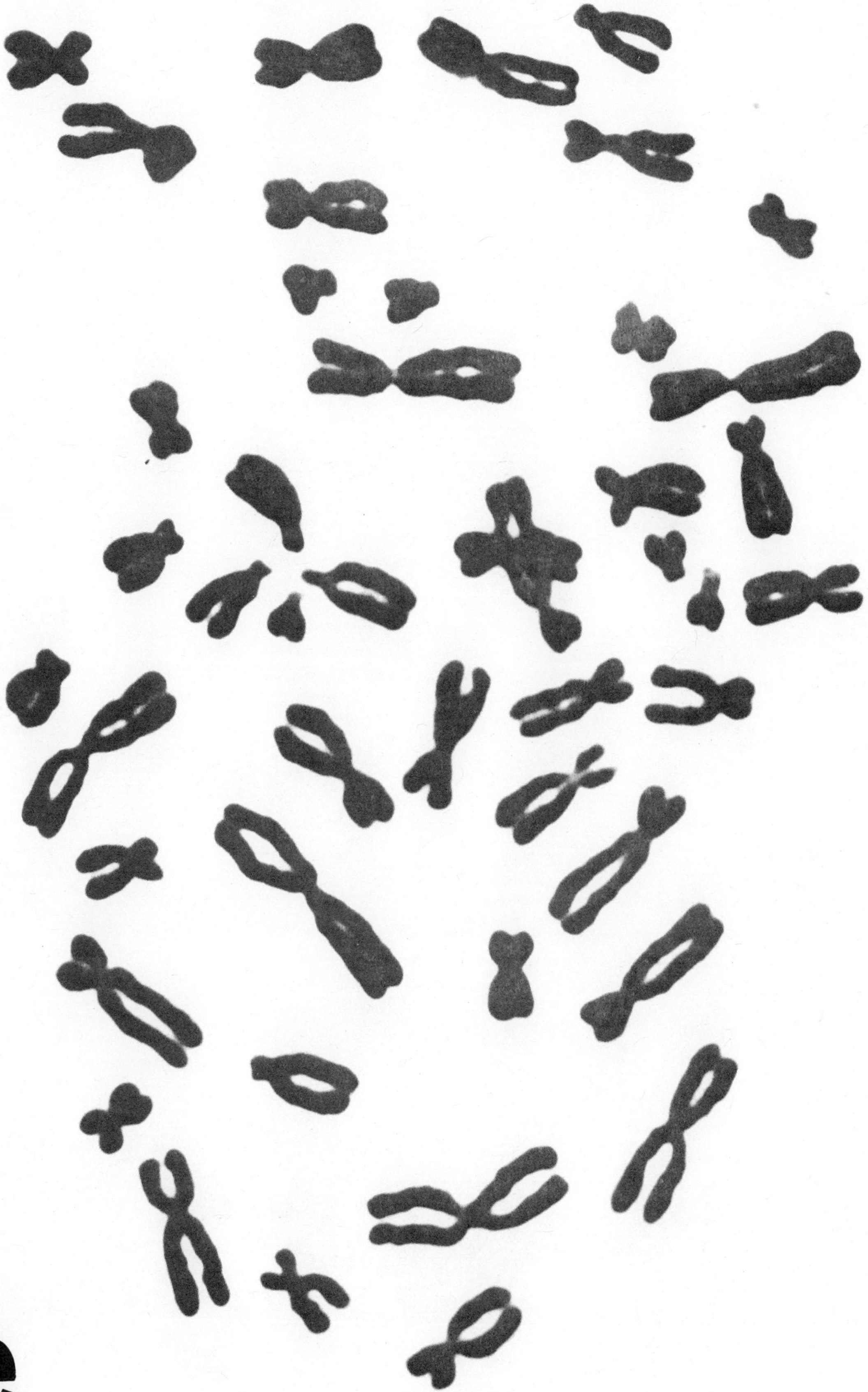

e

PROBLEM SET 11 (apropos sections 19-21)

1. What is transformation? What is the relevance of this phenomenon for genetics?

2. Sickle cell anemia is described as a molecular disease. Explain.

3. The genetic code is described as: (a) non-overlapping, (b) commaless, (c) degenerate, (d) universal. What is the meaning of each term?

4. What is a codon?

5. How do mutant hemoglobins verify the correctness of the genetic code?

6. What is the experimental rationale for recovering "nutritional" mutants in Neurospora?

7. In some kindreds of hemophilia A (classical hemophilia) the impaired blood coagulation is severe, in other kindreds the impaired blood coagulation is relatively mild. How do you explain this difference within the context of "inborne errors in metabolism?"

NOTES

A sample of answers to the problems

Problem 1-3. Dominant, autosomal trait; twins = dd and Dd.
Deaf twin transmits deafness to 1/2 progeny, normal twin to none.

Problem 2-2. $\frac{1}{4}$; $\frac{3}{4}$. albino = cc, normal = CC or Cc.

Problem 3-3. $\frac{1}{4} \times \frac{1}{4} = \frac{1}{16}$; $\frac{3}{4} \times \frac{3}{4} = \frac{9}{16}$

$10(1/4)^2(3/4)^3 = 270/1024$; $10(1/4)^3(3/4)^2 = 90/1024$

Problem 4-4. 1) $\sqrt{80/2000} = 0.2$, 2) $0.32 \times 2000 = 640$, 3) 0.16%, 4) 0.1667

Problem 5-5. $(\frac{0.025}{1.025})^2 = 0.000595$; $(\frac{0.025}{1.125})^2 = 0.000494$

Problem 6-1. a) 1/2; 1/2, b) 1/8; 3/8

Problem 7-4. a) middle gene = P; b) Xg-10-P-20-G

Problem 8-1. $r = 0.2$, $R_1 = 0.6$, $R_2 = 0.2$

predicted erythroblastosis = 3/2%

Problem 9-3. First child--exclude lover; second child--exclude father; third child--exclude neither.

Problem 10-b. Normal ♂

Problem 11-2. Anemia related to permanent change in β polypeptide of Hb molecule--at position 6 glutamic acid substituted for valine.